The Ostrich
Communal Nesting
System

MONOGRAPHS IN BEHAVIOR
AND ECOLOGY

Edited by John R. Krebs and
Tim Clutton-Brock

The Ostrich Communal Nesting System

BRIAN C. R. BERTRAM

Princeton University Press
Princeton New Jersey

Copyright © 1992 by Princeton University Press
Published by Princeton University Press, 41 William Street,
Princeton, New Jersey 08540

Library of Congress Cataloging-in-Publication Data
Bertram, Brian, C. R., 1944–
    The ostrich communal nesting system / Brian C.R.
    Bertram,
      p. cm. — (Monographs in behavior and ecology)
    includes bibliographical references and index.
    ISBN 0-691-08785-7
    1. Ostriches–Behavior. 2. Sexual behavior in animals.
    3. Ostriches–nests. 4. Social behavior in animals. I. Title.
    II. Series.
    QL696.S9B47   1992   92-22198
    598'.510456–dc20   CIP

This book has been composed in Lasercomp Times Roman
and Univers Med. 689

Printed in the UK

Editorial and production services
Fisher Duncan, 10 Barley Mow Passage,
Chiswick, London W4 4PH, UK

Contents

Preface

Surprisingly, this field study of ostriches grew directly out of my earlier research on lions in the Serengeti National Park in Tanzania. It was a chance encounter, while searching for lions, with an ostrich nest containing just recognizably different eggs, that prompted a much more thorough look: it appeared that contrary to what natural selection usually dictates, a female ostrich seemed to be doing all of another female's work for her. At that time, in 1973, there was a new and rapidly growing interest in cooperation and apparent altruism in animal social systems, and in how such behaviour could evolve through natural selection. Lion research was revealing the ways in which cooperation and altruism in that species could be selected for. In many respects, ostrich eggs promised to be more amenable subjects for comparative study than elusive mobile nocturnal carnivores.

Both the ostrich and its eggs have an improbable impressiveness about them. The close observer is struck by the size and the height of the eye, the power of the legs and the softness of the barbless feathers. The creamy white round egg, the equivalent of some 25 ordinary chickens' eggs, has a mystique which matches its food value. And as this study showed, the ostrich communal nesting system is complex, intriguing and adapted to the birds' difficult environment.

As with any study, it was not carried out in isolation, but depended on a great deal of help from others. I am indebted to three organizations for financial support. The study was conducted while I held a Senior Research Fellowship in the Research Centre at King's College, Cambridge from 1976 to 1979. Fortunately, the long university vacation was apparently designed to coincide with the ostrich breeding season in Kenya. Field work expenses were covered by research grants from the Royal Society and the Natural Environment Research Council – which means the UK taxpayer.

I am grateful to the Office of the President and the Wildlife Conservation and Management Department of the Government of Kenya for permission to carry out research work in Tsavo West National Park, and to the then Senior Park Wardens Ted Goss and Bill Woodley and their staff for friendly local assistance and advice. I was glad of association with the Zoology Department of Nairobi University and with the Department of Ornithology of the National Museums of Kenya.

The study benefited greatly from the ingenuity of Emrys Williams who produced excellent timing devices to operate cheap time-lapse cameras (Section 2.3), and of William Bertram who devised temperature-recording ostrich eggs (Section 2.4).

Continuous off-track driving in rough terrain can be hard on vehicles, so I am particularly grateful for the generosity of Rosemary and Alasdair Macdonald and of William Bertram for the loan of theirs.

Several people helped during the fieldwork in Tsavo, particularly Debbie Snelson and Anna Baird (now McCann) for the last field season; a good many others helped with observations and in other ways during shorter periods. Rosemary and Alasdair Macdonald and Ann and Mike Norton-Griffiths kindly provided invaluable bases in Mombasa and Nairobi.

A brief study visit to South Africa to look at ostrich farming practice was made successful through the kind help of Roy Siegfried, Alan and Valerie Burger, and several ostrich farmers at Oudtshoorn.

I am grateful for discussion and comments at various stages to many people, but particularly Debbie Snelson, Anna Baird, Tim Clutton-Brock, Dan Rubenstein, Steve Albon and Richard Wrangham. I am grateful too for the forbearance shown by many people over the years in not asking too often how the book was coming along. For the typing of the general result, I should like to thank Hazel Clarke, Mary Gillie, Elspeth Chaplin, Jacquie Cook, Alison Byard and Jennifer Owen. David Bygott's sketches have helped to bring the birds to life.

But the person who has contributed by far the most to the study has been Kate Bertram. That contribution has included three seasons of being a first-class joint ostrich observer and superb camp organizer, many hours of analysis of unexciting time-lapse photographs and general tolerance with her spouse throughout.

To all the above, I can only reiterate my grateful thanks.

1 The Ostrich

Running

1.1 The Bird Itself

The ostrich is well known to be the world's largest living bird. An adult male stands 2.1–2.7 m high (Cramp and Simmons, 1980). Females are generally a bit smaller. Information on measured weights of these birds in the wild is characteristically scarce; a small sample ($n = 13$) from Kenya averaged 111 kg, ranging from 86 to 145 kg (A.V. Milewski, pers. comm.).

Their size alone would render ostriches incapable of flight. Instead, they are adapted to a walking and running way of life. The very long bare legs carry the large bird economically (Fedak and Seeherman, 1979) and when necessary at up to 60–70 km/h (Brown *et al.*, 1982), with strides of up to 8.5 m (Smit, 1963). The reduction in the number of toes from the normal five to only two in the ostrich is probably another adaptation for fast running. The large inner toe (originally the third) and occasionally the smaller outer toe (originally the fourth) carry a powerful nail.

Ostriches' wings are considerably reduced. The feathers are without barbs, which contributes to their loose, soft appearance. The primary feathers are developed as large plumes, used in display and sought after by human beings. Much of the body is devoid of feathers, particularly the long neck, the whole of the legs and patches on the underside of the body. The adult male's body plumage is jet black, with the exception of white plumage on its wings and tail. The female's feathers are of a fairly uniform earthy pale brown-grey colour.

Plate 1.1   An ostrich head in close-up.

The eye is very large (50 mm across: Hurxthal, 1979), is protected by long eyelashes, and with good visual acuity it provides excellent vision from its high vantage point.

Unlike most birds, the male ostrich has a penis, which is everted when the bird defecates and urinates.

The eggs vary in size, weight (1100–1900 g) and shape (from ellipsoidal to almost spherical). The shells are creamy white in colour and about 2 mm thick.

The species name *Struthio camelus*, given by Linnaeus in 1758, derives from the Greek and Latin name *Struthocamelus*, by which the ostrich was known. The '*camelus*' is based on a supposed similarity to camels, probably in the strong fleshy feet, the prominent eye lashes, the large size and the desert habitat it can survive in.

The ostrich is the sole living species in the family Struthionidae. Half a dozen extinct species of *Struthio* have been described on the basis of fossil bone and eggshell fragments (Brodkorb, 1963). These species were mostly larger than the present-day species, and during the Pliocene and Pleistocene periods up to 5 million years ago, they occupied wide areas of China, India and Eastern Europe as well as Africa.

The higher classification of the ostrich has been considerably debated.

Plate 1.2   An adult male ostrich *S.c. massaicus* in Tsavo West National Park. The white tail is stained reddish brown by the local soil. A bolus of food is being swallowed.

It is generally considered (e.g. Cracraft, 1974, 1983) that the Rheas (*Rhea* and *Pterocnemia*) from South America are their closest relatives, also in the family Struthionidae. More distant are the Cassowaries (*Casuarius*) from New Guinea to Australia and the Emu (*Dromaius*) from Australia. All of these large, flightless birds, along with the New Zealand Kiwis (*Apteryx*) and some other extinct giant birds including the Moas (*Dinornis*) of New Zealand and the Elephant Birds (*Aepyornis*) of Madagascar, are classified together as the Ratites. Their relationship to other birds is not clearly understood, but they are generally considered to have descended from a common ancestor (Sibley and Ahlquist, 1981), which was capable of flying (Bruning, 1991).

1.2   Ostrich Subspecies

During the twentieth century, five subspecies of ostrich have been validly recognized (Brown *et al.*, 1982). The Masai Ostrich, *Struthio camelus massaicus* Neumann, forms the subject of this book and so is dealt with first. As the name suggests, it is found in East Africa, from central Kenya to south-west Tanzania (see Fig. 1.1). The bare skin of the male's neck is pink, and becomes red during the breeding season; his thighs likewise, are

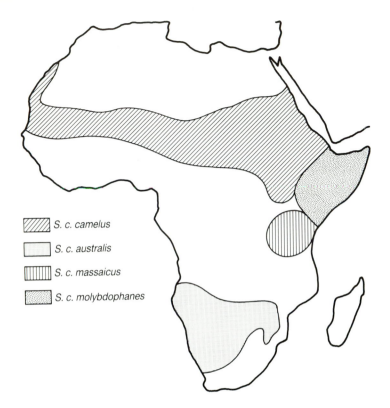

Figure 1.1   Distribution map of ostrich subspecies in Africa. Information derived from Brown *et al.* (1982) and Mackworth-Praed and Grant (1952, 1962).

bright pink in colour. The white tail feathers are almost invariably soiled and so usually brownish or reddish in colour. There is a small ring of white feathers as the black body feathers give way to bare skin about a third of the way up the neck.

Adjacent to the Masai Ostrich to the north-east is the Somali Ostrich, *Struthio camelus molybdophanes* Reichenow, from the horn of Africa. It inhabits north-east Kenya and south-eastern Ethiopia and Somalia. It is distinguishable from the Masai Ostrich particularly by the male's deep grey-blue neck and legs, bright red tarsal scales, and because he has no white neck ring. His tail feathers are usually unsoiled and therefore bright white. Also, Somali Ostriches have a bare crown patch, a grey rather than a brown iris, and they usually hold their beaks a little higher (Hurxthal, 1979).

The South African Ostrich, *Struthio camelus australis* Gurney, occurs

Plate 1.3   A male ostrich of the Somali race. The neck and legs are blue-grey, and the white wing and tail feathers fluffier and apparently less prone to stain.

south of the Zambezi and Cunene Rivers. It is another blue-necked race, also with red tarsal scales but with no bare patch on the crown. It is mainly this race, hybridized with others, which has been domesticated on ostrich farms (see Section 1.4).

The North African Ostrich, *Struthio camelus camelus* Linnaeus, is distributed across the southern Sahara and Sahel region from Mauritania to Ethiopia. This is the tallest race, and the most similar to the Masai Ostrich in having a bright pink neck and thighs; but the bill and tarsus are redder, particularly in the breeding season, and there is a bare crown patch.

A so-called Dwarf Ostrich, *S. c. spatzi*, from the Rio de Oro, described on the basis of a small zoo specimen and eggshell fragments, is no longer recognized as being a valid race (Cramp and Simmons, 1980). The Arabian Ostrich. *S. c. syriacus* Rothschild, which once inhabited Arabia and neighbouring countries, was wiped out between 1940 and 1970.

Wild ostrich populations everywhere are declining, their ranges are shrinking, and the survival of some subspecies is severely threatened. The North African Ostrich in particular which once occupied a vast range, has totally disappeared from the northern side of the Sahara (Egypt, Libya and Tunisia) within the past 150 years, and is now almost extinct in the western Sahara. Its numbers are no longer high anywhere, as the species is vulnerable to disturbance, delinquent shooting and human plundering of nests.

The Masai and Somali Ostrich races are holding their own in the good wildlife areas within their range; but as human numbers in their habitat rise, ostriches become increasingly vulnerable to disturbance and casual plundering of their nests. The South African Ostrich has been eliminated from most of its former large range, mainly due to conversion of its habitat for farming and ranching activities. It survives now in the more arid parts of its range, particularly in Namibia and Botswana.

The only subspecies which are nowadays contiguous are the Masai and Somali Ostriches in parts of Kenya where they overlap for a few dozen kilometres. No naturally occurring hybrids have ever been reported there, although a limited amount of interbreeding did take place in the Nairobi National Park following the release into the local Masai Ostrich population of a couple of male Somali Ostriches. There are differences in the courtship displays of the two races (pers. obs.; Hurxthal, 1979) and these may prevent hybridization except in artificial situations such as the above and on ostrich farms.

Despite the declining status of some races, the ostrich is not yet listed in the Appendices of the CITES (Control of International Trade in Endangered Species) Convention; this means that ostriches or their products taken from the wild can still legally be taken across most international borders. Domestic legislation in a number of countries provides a degree of protection against human activities. In Britain, for example, ostriches are classed in Group A1 of the listing of the Department of the Environment, and import licences are required to move ostrich eggs into or out of the country.

1.3 Habitat and Feeding

Ostriches may be found in a variety of open habitat types (Brown *et al.*, 1982). They avoid areas of thick bush or of heavy tree cover, but inhabit wooded grasslands and other more open country. Semi-arid, open, short-grass plain is usually associated with the highest ostrich densities. Highly productive and very well vegetated areas usually support high populations of competing herbivores, and therefore high predator populations. Adult ostriches are vulnerable to predation particularly by lions (*Panthera leo*) if there is a great deal of cover. And ostrich nests are vulnerable to nest predators such as hyaenas (*Crocuta* and *Hyaena*) for about 2 months; a very high hyaena density in an area could result in almost all ostrich nests being discovered and destroyed before hatching takes place.

At the other extreme, ostriches are able to thrive in very poorly vegetated areas, and can be found in dry savannah to desert. They exhibit a number of adaptations for dealing with the harsh desert conditions – extreme temperatures, a lack of water and little food.

Ostriches are tolerant of high temperatures. They rarely seek shade, as most desert antelopes regularly do. Their feathers are excellent insulators, minimizing heat gain due to direct solar radiation, as well as reducing heat loss during cold desert nights. Temperature control is achieved by erecting and flattening the body feathers (Louw et al., 1969) and by panting at very high temperatures (Crawford and Schmidt-Nielsen, 1967).

The bulk of ostriches' water needs can usually be obtained from their food, achieved partly by their feeding early in the morning, particularly on hygroscopic plants with a high moisture content (Louw, 1972). If free water is available, the birds will make use of it, and under certain conditions they will trek long distances to obtain it (Sauer and Sauer, 1966a). Generally, their distribution is almost independent of water as they can withstand a considerable degree of dehydration (Cloudsley-Thompson and Mohamed, 1967). Their urine consists of uric acid carried in mucus, so minimizing water loss (Louw et al., 1969).

A desert environment has sparse food items, and ostriches are well adapted to harvesting them. The bird's excellent eyesight detects the food item at a distance, and the beak at the end of the long, light, mobile neck can gather the food delicately and economically.

Ostriches eat a variety of species and parts of plants, particularly dicotyledonous ones. The plants eaten include succulents, herbs, shrubs, grasses, creepers and bushes (Brown et al., 1982); leaves, flowers, small fruits, seeds and seed pods or heads are all selected and consumed. In captivity, ostriches are considered omnivorous, as they relish meat and will pick up almost any small bright object. In the wild, however, animal protein (in the form of lizards, locusts, termites, etc.) clearly forms only a minute proportion of their diet and they appear not to go out of their way to obtain it.

The food selected is evidently of relatively high quality, and ostriches thrive and grow quickly on it. They can lay down quantities of subcutaneous fat reserves against times of food shortage.

1.4   Relationship with Man

Ostriches have had a long and varied relationship with man. They feature

Plate 1.4 Two female ostriches feeding from the sparsely scattered nutritious food items in Tsavo West National Park.

in the folklore and carvings of the Kalahari Bushmen (Bleek and Lloyd, 1911), appear as paintings and carvings in caves in the Sahara dating from 5000–10 000 years B.C., and were considered holy by the Assyrians (Smit, 1963). The Bible (Leviticus 11:13 and Deuteronomy 14:12) prohibited the eating of ostriches along with a whole range of other birds, apparently as being unclean. Elsewhere in the Bible (Job 39:17), referring to the ostrich, it is stated that 'God hath deprived her of wisdom, neither hath he imparted to her understanding'.

The belief in the bird's stupidity persists in the still current notion that ostriches bury their heads in the sand in face of danger. No-one has witnessed this at first hand. There are several ways in which the story could have originated: a feeding ostrich may have its head concealed among low vegetation for considerable periods; the small neck and head of a very distant ostrich can be invisible; and a nesting ostrich when approached by humans or other predators will often stretch its neck and head flat out along the ground where it is not easily noticed.

Ostrich eggs have long been prized. The Bushmen, and later European sailors, found them an invaluable food source which kept fresh for long periods. The Hottentots used complete empty shells as practical water vessels (De Mosenthal and Harting, 1879). As carved vases and cups, ostrich eggshells were being used in Mesopotamia 5000 years ago (Laufer, 1926), as well as in Ancient Egypt, Carthage and post-Renaissance Europe. The holy properties of the shells were used to help and protect Ethiopian Coptic churches (Brown *et al.*, 1982) and buried Phoenicians. And shell fragments have been extensively made into necklace beads by Bushmen and others.

People have also made use of the birds themselves for many centuries. The Struthophagi tribe in Egypt used to hunt ostriches by stalking them disguised in an ostrich skin (De Mosenthal and Harting, 1879), as did the Kalahari Bushmen (Sauer, 1971). (The Australian Aborigines also hunted emus in this manner.) In Tutankhamen's tomb, there is a depiction of him hunting ostriches by bow and arrow from his chariot, apparently a privilege of the Pharoahs. The Arabs also used to hunt them on horseback. Guns have nowadays made ostrich hunting much easier.

Ostriches have been hunted for a variety of reasons, and most parts of a killed bird are put to use. The skin, being flexible but tough, has been used for making protective jackets worn in the Arab world. The mad Roman emperor Heliogabalus of the second century A.D. had 600 ostrich brains served at a banquet, but in general the meat has not been specially sought after. A large number of ostriches were shot in South Africa following the discovery of diamonds in the gizzard of one of them (Sauer, 1971). But their feathers have been the birds' main attraction.

Ostrich feather plumes have been used for decoration for many generations. They are common in the headdresses of African warriors. They adorned the fans of Assyrian Kings and of Popes, the horses of Tutankhamen's chariot, and the headdresses of Greek, Roman and Turkish generals. The ostrich plume frequently featured in the hieroglyphics of Ancient Egypt, where with its symmetrical appearance it was adopted as the symbol of justice and truth (Smit, 1963).

The European demand for ostrich feathers for personal adornment began in the fourteenth century; it grew steadily, and resulted in widespread persecution of the species. Without domestication, ostriches might well have been completely exterminated by now. For centuries, a few people had captured odd young birds as chicks, tamed and reared them, and kept them as pets, or for riding as in ancient Rome and Egypt (Smit, 1963). But the farming and breeding of them in captivity only commenced in about 1863 in South Africa.

The ostrich farming industry was phenomenally successful, fuelled by prices of £5 and more per pound of feathers. The development of husbandry techniques, and particularly of ostrich egg incubators, resulted in an increase of the captive population from 80 birds in 1865 to over 150 000 within 20 years (Smit, 1963). Large fortunes were made. Many

Plate 1.5   Ostrich hunting by bushmen in the nineteenth century.

ostrich farms sprang up, mainly in the Little Karoo region centred on Oudtshoorn in the Cape Province of South Africa, but also elsewhere in South Africa, Kenya, Egypt, Australia, New Zealand, the USA and Argentina. At the peak around 1913, there must have been at least 1 million birds in captivity.

Careful selection of breeding birds was carried out, studbooks were kept, different strains were recognized, and ostriches from elsewhere in Africa were introduced. As a result, the present-day domestic stock, although derived mainly from the South African Ostrich, has an admixture of North African Ostrich and Arabian Ostrich blood. Improvements in diet, treatment of parasites and understanding the causes of growth defects in feathers all resulted in improvements in feather quality.

From being South Africa's fourth largest export earner in 1913, the ostrich feather market crashed during the economic depression after the Great War, helped by the advent of the motor car (too low inside and too fast outside for large feathered hats). Bankruptcies abounded. The industry survived on a much smaller scale in Oudtshoorn, and is still thriving there but diversified with other types of agriculture (personal observations and enquiries).

Today, ostriches on the farms in Oudtshoorn are kept mainly as breeding pairs in small enclosures of about 0.25–1.0 ha. They are fed largely on lucerne growing in their irrigated paddocks, plus a little maize. They lay about three clutches per year. Some of their eggs they are allowed to hatch themselves, whereas others are put in incubators (which are less efficient). Incubator-hatched chicks are returned to adult pairs for fostering and rearing. Non-breeding birds are kept in large groups in very large enclosures.

At about 2 years old, inferior birds are slaughtered (about 30 000 per year) and converted into skin for tanning, dried biltong from the leg meat, meat-meal and bone-meal. The superior birds are kept for about 15 years, being plucked for their feathers every 6–9 months. Breeding birds may be kept for 30 years or more.

The main earnings (about 54%) of the industry nowadays still come from the feathers, followed by the skin (31%) and meat (13%) (Oudtshoorn Museum's figures). The best white plumes will sell for about £3 each. They go to organizations as diverse as fashion houses and the *Folies Bergères*. Boas are made for the luxury market, and feather dusters on a massive scale. The skin goes to make purses, handbags and the like. Eggshells, decorated or plain, are exported in numbers. Tourists to a few of the ostrich farms bring in a substantial revenue, buying the above and other souvenirs (eggshell ashtrays, ostrich foot lampstands, etc.), and paying to observe the intricacies of this fascinating industry.

Plate 1.6   A male ostrich acting as foster parent to a brood of chicks at an ostrich farm at Oudtshoorn in South Africa.

1.5   Previous Ostrich Studies

The anatomy and taxonomy of ostriches and the other ratites received a great deal of attention during the past 100 years. The state of the fossil record of ostriches has been summarized by Brodkorb (1963), and this evidence and other published material relevant to ratite classification has been reviewed by Cracraft (1974).

There is a very extensive literature on ostrich farming. De Mosenthal and Harting (1879) provided a good early account, and Smit (1963) a retrospective summary. Professor J. E. Duerden published over 40 papers between 1907 and 1920, mostly in the *Agricultural Journal of the Union of South Africa* and its predecessor, on a range of biological topics related to the ostrich industry, including a quantity of experimental work. Mitchell (1960) produced a bibliography of over 300 titles dealing mainly with the ostrich farming aspects of physiology, diseases, husbandry, feathers, skins, marketing and official documentation, but also including the few anecdotal accounts of these birds in the wild.

Until recently, very few systematic studies had been made of wild ostriches. The reasons are partly conceptual – large vertebrate behavioural

Plate 1.7   An ostrich egg incubator at Oudtshoorn with newly hatched chicks.

field studies really only got under way some 25 years ago – and partly practical: "the difficulties of getting information concerning a shy bird endowed with almost miraculous eyesight, and which can out-distance fast mechanical transport, can easily be imagined" (Bannerman, 1930).

The first field studies of ostrich behaviour and ecology were made by Franz and Eleonore Sauer during research visits to Namibia in 1957–58, 1964 and 1969–1970. In a series of publications between 1959 and 1972, they described the social and reproductive behaviour patterns of wild South African Ostriches with a group-laying harem system, but presented relatively little of the quantitative data they collected. Jarvis *et al.* (1985) collected breeding data for this race in Zimbabwe.

Leuthold (1970, 1977) monitored one nest of the Somali Ostrich, and analysed breeding records for this race in Kenya. On the Masai Ostrich, Adamson (1964) made observations on production in one nest.

Lew Hurxthal carried out a PhD study of the Masai Ostrich, making observations mainly during 1971 and 1972 in Nairobi National Park, and producing his thesis in 1979. That provided data at the population level on reproduction and mortality, emphasized synchrony, categorized the variety of signals used in communication, outlined the different ranging patterns of male and female, described the detailed behaviour of nesting, emphasized the degree of mate choice and of communal laying, analysed the process of merging of broods of chicks, and discussed the role of size and of the different sexual strategies in the evolution of the ostrich breeding system. (It is impossible to summarize adequately a 270-page thesis, plus tables and figures, in a single paragraph.)

The above represent the bulk of what has been published on the behaviour and breeding of wild ostriches. Some other data on wild Masai Ostriches have been provided incidentally in the course of two other types of studies. Ostrich distributions have been published from wildlife distri- bution studies such as those of Cobb (1976), Foster and Coe (1968) and Western (1973) in Kenya. Predation on Masai Ostriches has been recorded in the predator studies of Kruuk (1972) and Schaller (1972), and on their eggs in many studies including those on Egyptian vultures (*Neophron percnopterus*) by Van Lawick-Goodall (1968), Boswall (1977), Brooke (1979) and Thouless *et al.* (1989).

Finally, information has been gathered on aspects of the physiology and adaptations of wild (Siegfried and Frost, 1974) and captive ostriches (Cloudsley-Thompson and Mohamed, 1967; Crawford and Schmidt-Nielsen, 1967; Fedack and Seeherman 1979; Louw *et al.*, 1969).

1.6   Summary of Ostrich Behaviour

Ostriches are totally diurnal. They are on their feet for most of the daylight hours, except when dustbathing, resting or nesting. They invariably sit down at dusk and remain inactive throughout the night unless disturbed.

The species is loosely gregarious. An ostrich is often on its own, but is attracted to others, for short periods at least. Most encounters with ostriches are with very small groups, but they may in places aggregate in very large numbers, even hundreds (Sauer and Sauer, 1966a). The chicks and juveniles are strictly gregarious, and always remain in compact groups. There appear otherwise to be no long-term bonds.

Ostriches are great travellers, and for much of the day are in motion, meandering as they feed opportunistically and often casually, striding purposefully towards other individuals or nest or drinking sites, loping steadily when involved in intraspecific pursuits, and sprinting and zigzagging when surprised, frightened or excited. Apart from locomotion, the legs are used for making scrapes in the ground, for striking opponents and small predators, and for scratching the head.

The wings are used for a variety of display purposes in addition to their function in controlling the bird's temperature, in counterbalancing during fast manoeuvring, in protecting eggs and young, and in whisking away flies. Ostriches use a repertoire of visual displays fully described by Hurxthal (1979), and the wings play a major part in many of them. The contrast of the white wing primary feathers against the black body feathers makes such displays particularly conspicuous in males. The main displays involving the wings are as follows:

Kantling – by males towards females to induce readiness to copulate, and towards other males in agonistic encounters. While squatting, the male waves his quivering spread wings alternately in the air as he rocks from side to side. The same movements are made during copulation itself.

Full threat – by a male just before attacking, and just before copulation. Both wings are raised high above the body, the primaries being well spread.

Wing-flagging – by a male in mildly aggressive encounters. Alternate wings are flicked up and down beside the body.

Soliciting – by submissive and possibly receptive females towards males, and by males when showing a nest site to females. Both wings are spread, lowered and quivered intermittently, while the head is held low but jerked up at intervals.

Distraction – by males and females with chicks (or nests near hatching), towards predators including human beings and vehicles. The spread wings are held low and rowed erratically in unison, while the neck waves unpredictably and the bird may periodically collapse.

The general posture of the body communicates information to other individuals. In general, a more confident and aggressive bird holds its head and neck high, with the front of the body tilted upwards, and the tail up. By contrast, a submissive individual conveys the fact by holding its head low and its tail down.

The bare skin of the male's neck changes from whitish to bright pink as the bird comes into breeding condition and becomes territorial. In

Plate 1.8 · A male ostrich scuttles away (from newly hatched chicks), performing a predator distraction display.

addition, a flush of colour can make the neck noticeably brighter during aggression or courtship. The red neck is particularly conspicuous when a male ostrich is performing its booming song, as the neck then inflates to at least double its normal thickness. The sound, a deep and powerful 'mwoo-mwoo-mwooooo', can be heard up to 1 km away, but is not given very frequently. Territorial males sing when alone, when approaching a hen to court her and in aggressive interactions with other males.

Other vocal signals are few. A coarse hiss sometimes accompanies the open-beak threat of one bird towards another at close range. A snapping of the beak can be heard and seen in mild defensive threat situations.

Ostriches preen their feathers frequently and indelicately. Their necks and wings are usually in motion as the birds snap at and swish away flies.

1.7   This Study

The main aim of this study was to investigate the communal nesting system of ostriches. It was known previously that the species was a cooperative breeder in the sense that most nests contained the eggs of more than one female. In recent years, there has been an increasing interest in cooperative breeding, and more and more species of birds have been found to have more complex social systems than had previously been thought.

With ostriches, it was suspected that only one female was carrying out the work of incubating the eggs of several females. Such apparent altruism in caring for the offspring of other individuals is relatively rare in nature. It deserves investigation when it does occur, because at first sight it appears improbable that it should be favoured by natural selection. After all, if a bird devotes some of its energies to the raising of the offspring of its neighbour, it would be expected to produce relatively fewer of its own offspring compared with that neighbour. Such behaviour should put the altruistic individual at a selective disadvantage. We know that nest parasites such as cuckoos reduce the reproductive output of their unfortunate hosts, and have caused the evolution of a variety of defence mechanisms. Intra-specific nest parasitism is much harder to detect, and less well known. It is also a more complex phenomenon, and the relative advantages and disadvantages are not necessarily as clear. This study was designed there-fore to investigate the phenomenon, to discover the extent to which birds laid in one another's nests, to find out the reproductive consequences of laying eggs in different nests, and to try to determine how the communal nesting system is maintained by natural selection.

The fieldwork was carried out in Kenya in 1977 (13 July to 4 October), 1978 (12 July to 24 October) and 1979 (11 July to 25 September). A previous visit in 1976, combined with the advice of a number of people (especially L.M. Hurxthal, P. Lack, S.M. Cobb, A. Root and P.H. Hamilton) and published material (Leuthold, 1970), made it clear that the Masai Ostrich in Kenya exhibited a single annual breeding season during the second half of the year. The southern part of Tsavo West National Park was selected as a study area for several reasons: the main laying time was likely to occur in the middle of the dry season study periods; com-munications were fairly good; visibility was reasonable; disturbance by human beings was likely to be minimal; and the terrain allowed cross-country driving. As a bonus, our campsite beside Lake Jipe with views of Mount Kilimanjaro was an idyllic spot with a superb diversity of fauna.

The work in the first two years was carried out with the assistance of my wife Kate Bertram; in the third year we had also the services of two field assistants, Anna Baird (now McCann) and Deborah Snelson, in a second vehicle, enabling independent observations to be made.

Additional information was obtained in the course of a 3-week study visit to ostrich farms at Oudtshoorn in South Africa where more intensive observations of incubation behaviour were possible than with wild birds. Data from one nest observed in the Serengeti National Park in 1973 were useful in stimulating the study and in providing a little valuable extra information in Chapters 4 and 6.

In writing up the results of the study, I have tried first to describe the social and ecological context behind the ostrich nesting system, then to set out the reproductive consequences of the different breeding strategies used by different individuals, and to quantify them in so far as is possible, and finally to consider how the ostrich communal nesting system is maintained, and how it probably evolved.

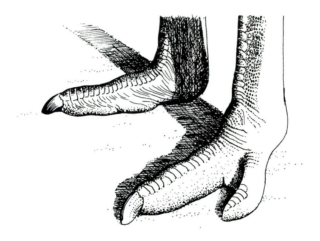

Feet of ostrich

2 Methods

Observing ostriches

In order to understand in their context the data presented in subsequent chapters, a summary is first provided in this chapter of the environment in which ostriches were studied in Kenya. The methods by which the birds could be observed and recognized individually are outlined, and the kinds of observations that were possible given the birds' timidity. One of the main limiting factors in the study was the difficulty of finding and observing surviving ostrich nests, and a variety of indirect methods proved to be necessary.

2.1 The Study Area

The fauna and flora of Tsavo West National Park have been described by Cobb (1976). The main study area chosen for this study was at the south-west corner of the National Park, close to Lake Jipe on the Kenya–Tanzania border. Observations of ostriches and their nests took place within an area 45 × 20 km, with the majority being concentrated within an intensive study area 17 × 10 km (see map, Fig. 2.1).

The intensive study area consists of gently undulating savannah. The sides and tops of the undulations are profusely dotted with bushes and with less frequent *Acacia*, *Commiphora* and baobab trees. The shallow drainage lines in general are less bush-covered and some of them are wide open grass areas. The soils are mainly hard and red on the ridges, blacker and softer in the drainage lines.

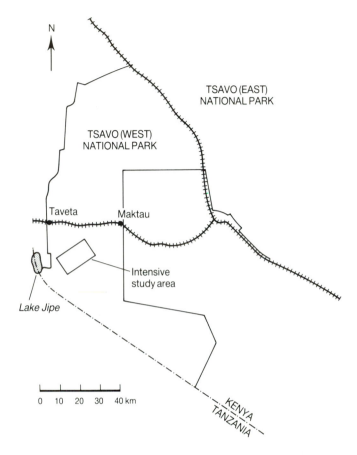

Figure 2.1   Map of the Lake Jipe region of Tsavo West National Park in Kenya, showing the intensive study area.

Very little rain generally fell between the months of June and October. The average annual rainfall pattern is shown in Fig. 2.2, based on data collected at the National Park Headquarters at Kamboyo, approximately 80 km north of the intensive study area.

The extent of the ground vegetation cover varied considerably from one study season to the next, a reflection partly of differences in rainfall between the years. At the start of the 1977 season, the country was very dry, grass cover was dried and sparse, most bushes were without leaves, and visibility in general was good; these conditions prevailed and intensified throughout the study period of that year.

At the start of the 1978 season, as a result of a much wetter preceding

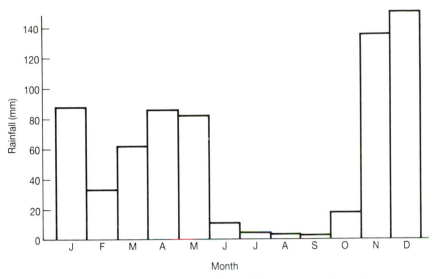

Figure 2.2 Average monthly rainfall in Tsavo West National Park (measured at Kamboyo 1953–83).

rainy season, the grass cover was much thicker and less dry, most of the bushes still had their leaves, and the atmosphere was generally hazy; thus observation conditions were poor. Extensive grass fires during the study period in 1978 removed the grass cover from about half of the intensive study area, resulting in improved observation conditions. The 1979 season was intermediate between the two previous years, in terms of amount of vegetation cover and extent of burning.

The vertebrate fauna of the area is rich. Elephants (*Loxodonta africana*) and giraffes (*Giraffa camelopardalis*) are abundant, as are zebra (*Equus burchelli*), impala (*Aepyceros melampus*) and Coke's hartebeest (*Alcelaphus buselaphus*). Black rhino (*Diceros bicornis*), frequent during 1977, had been almost eliminated by poaching by 1978.

Among the large carnivores, lions (*Panthera leo*) and leopards (*Panthera pardus*) were observed quite frequently, and cheetahs (*Acinonyx jubatus*) occasionally. Spotted hyaenas (*Crocuta crocuta*) were heard occasionally but almost never seen, and were probably scarce. Black-backed jackals (*Canis mesomelas*) were abundant; they were observed much more often in 1979 than in the previous two years, and possibly were more abundant as a result of a Nile Rat (*Arvicanthis sp.*) 'plague' in 1978, itself probably a consequence of high rainfall and abundance of seeds.

2.2   The Study of Birds

a) *Detection.* All observations of live ostriches were made from a vehicle. Most preliminary searching was done while driving slowly along roads, tracks or previously navigated routes, stopping at intervals to scan the countryside with binoculars. Surprising as it may seem, ostriches are often difficult to detect. Although the black plumage of the males makes them relatively conspicuous, the earthy brown colour of the females makes a motionless distant bird easy to overlook. In addition, bush cover means that blind spots actually take up a significant proportion of the landscape. As shown later (Section 3.2), single ostriches were generally detected at distances averaging over 900 m for males and 500 m for females. Groups were usually detected from further away than single birds, particularly if one or more of the members was male. The distance of detection also varied with observation conditions, landscape and with the frequency of stops to scan as opposed to searching while travelling.

The difficulties of detecting birds resulted in a significant proportion of time being spent in trying to do so, and meant that individuals could not be found at will. Radio-tracking, which would have enabled this, was not used because of the vast problems of catching timid adult birds to attach radio transmitters.

b) *Observations.* At each sighting of an ostrich, the following were recorded:

- Date.
- Time.
- Location – by description and measured distances from known land-marks.
- Group size and composition – males, females, immatures and juveniles, within 150 m of one another.
- Detection distances – how far from the observer when first detected.
- Method of detection – whether by eye when travelling or by binocular scan.
- Which bird in the group was detected first.
- Duration and distance of observation.
- Activity of the birds (see below for categories).
- Which individual birds if known, or descriptions of characteristics if not known.

In some cases, birds seen were observed for prolonged periods; on other occasions, the group was watched only for the odd minute. Many of the sightings of ostriches were made incidentally in the course of our travelling for other purposes such as when on the way to known nests, or when going to or from the camp, or when searching for different individuals. The problems of finding nests (Section 2.3) were so great, and the rate of nest destruction so high (Section 5.3), that in the first two seasons relatively little time could be spent on lengthy systematic observations of known individuals. Only in the 1979 season, with the help of field assistants and a second vehicle, could time be set aside for such continuous watches without sacrificing nest data.

During the continuous watches of known individuals, birds were observed through binoculars, from a vehicle, at distances of between 100 and 500 m. The birds' timidity made attempts to approach closer undesirable. At regular 5-min intervals, the observer classified the birds' behaviour (during the next 15 s) into one of the following mutually exclusive categories:

- Alert – standing still, with the head up.
- Walking.
- Feeding – head down among the vegetation, and the bird taking relatively few steps.
- Nibbling – feeding quite frequently while walking fairly steadily.
- Running.
- Preening.
- Sitting down on the ground.
- Dustbathing.
- Soliciting – by females.
- Sexual behaviour – courtship, mating and kantling.
- 'Eggery' – activity when standing at the nest, often with head down among the eggs.

At every sixth such observation, i.e. at 30-min intervals, the observer also recorded the following:

- Distance travelled during the preceding 30 min.
- Number of other ostriches within 150 m, and their identity.
- Number of instances of mating, booming, kantling, chasing or being chased.

The length of any continuous watch on a particular individual varied. It depended in particular on the time of day that the bird was discovered (which determined how long until dusk), on whether the bird went to sit at a nest (after which the watch was discontinued), on whether contact with the bird was lost because of the difficulties of following fast-moving birds cross-country through difficult terrain, and on what other calls there were on the observer's time that day. The 93 continuous watches averaged 4 h 41 min in duration.

Opportunistic observations were made of male courtship behaviour, mating and nest-site display behaviour whenever it occurred.

c) Individual recognition. Where possible, individual birds were identified, using natural features. It is presumed that there is as much individual variation in ostriches as in other vertebrates: the main problem when studying ostriches in the field is to get close enough to the relatively timid birds to observe the small characteristics by which individuals differ.

Features used to distinguish individuals included scars, injuries, feather colours, and presence and position of feathers on bare skin areas, as well as general features of sex and size. Some features enabled us to recognize individuals with absolute confidence over long periods, such as female Bwing, a small uniformly grey bird with a dislocated left wing, observed in all three seasons. Some features, such as a distinctive pale or broken feather, enabled recognition of individuals within a season but not between successive seasons.

In general, males were easier to recognize than females for three reasons. First, the presence of the odd small black feather contrasting with bare pink thighs provided a relatively conspicuous marker, from behind (as we were if we approached too close). Second, the males that were territorial during the study periods tended to be in the same smaller areas; as a result, they were more often observed and so became less timid and more observable from close quarters. Third, there were fewer individuals to choose between.

The number of known individuals increased over the three study periods, partly because of our greater expertise, partly because there were twice as many observers, and partly because the birds gradually became somewhat less timid. In the final year, known individuals numbered 14 adult males and 24 females (Section 3.2).

For ease of communication between the observers, individual ostriches were given names which were determined either on descriptive grounds (Brokenwing, Lame, Dirtytail, Freak, Streak, Multi-feathered Leg, Goldie, Eartha, etc.), or for largely frivolous reasons (Adolphus, Josephine, Nemo, Stalin, Sista, Philippa, etc.). For convenience in this book, some of these names have been abbreviated as appropriate.

Plate 2.1   A female ostrich head in close-up. The small protuberance on her neck, probably a healed cut, would make her an easily recognizable individual.

2.3 The Study of Nests

The study of nests was somewhat piecemeal because they were discovered at many different stages. In addition, most were destroyed by predators after surviving for different and unpredictable periods of time.

a) Finding nests. Just finding nests was the greatest problem of the whole study. Ostriches are shy and keen-sighted birds, and are very wary in their behaviour if they can detect potential danger such as human observers,

Plate 2.2 A male and female close together, with the male soliciting with lowered wings quivering. This would be a strong indication that there was a scrape or early nest there.

even distant ones. Doubtless, natural selection has favoured such wariness.

Nests were found by the following methods:

1. *Flying*. Eggs in nests, although invisible to the terrestrial observer except within perhaps 10 m because of the concealing ground vegetation, can be conspicuous from above. However, aerial searches were prohibitively expensive, largely impractical for us and rarely successful.

2. *Seeing a male showing a site to a female*. Occasionally, a female following a steadily walking male was being led by him either to an early-stage nest or to a shallow scrape which might eventually become a nest. At such a location, the male lowered both his wings and quivered them – the soliciting display. The behaviour was not very frequently seen, and most of the scrapes dug and shown to females never became nests.

3. *Seeing one or more females first standing at a particular spot and then briefly sitting (to lay)*. Females are particularly wary when approaching nests. A continuous watch on an individual female only occasionally resulted in a nest being found, because there was no way of knowing whether or not she was currently laying, and if she was, whether or not the day in question was her laying day (because ostriches lay on alternate days: see Section 4.2).

4. *Following a male to a nest before dusk*. If a male was suspected of

owning a late-stage nest, it could sometimes be found by keeping a circumspect distant watch, where possible, of where he went before dusk. Nests found in this way were the few relatively late-stage ones still surviving.

5. *Chance encounter*. A few nests were found by chance in the course of driving over the landscape. This was more likely with destroyed nests where scattered pieces of white eggshell made them more conspicuous.

In total, 56 nests were discovered using these five methods. Obviously, they were found at different stages: only five as scrapes before laying had started, 35 during the laying stage, eight during the incubation stage, and eight only after the nest had hatched or been destroyed.

b) *Observing nests*. The behaviour of ostriches at nests was observed in three ways:

1. *Binoculars from a vehicle*. This was only possible from great distances from the nest (over 400 m), or the birds were unduly disturbed and often failed to approach the nest. The vehicle would be parked so that it was concealed behind a bush as much as possible. At such ranges, individuals could rarely be recognized until after they had left the nest (when we could approach them in the vehicle for identification), but the general comings and goings of different birds in the vicinity, and their courtship and layings, could be observed. Nothing could be seen of the eggs.

2. *From a hide*. Limited use was made of a small portable grass-covered hide which could be erected at a distance of 20–30 m from a nest. Its disadvantages were a restricted view of anything going on except right at the nest, a lack of mobility to follow birds, personal risks from aggressive elephants, and a continued inability to observe the eggs in the nest.

3. *Time-lapse photography*. We made extensive use of time-lapse photography. Cheap Sankyo ES-33 Super-8 cine-cameras with single-frame exposure capability were linked to cheap battery-operated variable timing devices manufactured by Emrys Williams of the Cambridge University Engineering Department. These could take pictures at rates between 1 per second and 1 per 30 min – in the main, they were used at 1 frame per minute. Fitted with a photo-electric cell that switched the device off at night, a Super-8 cassette lasted just over 4 days. The system was small enough that it could be disguised and attached inconspicuously to a branch situated or positioned about 15 m from the nest, and ignored by the birds. It was also expendable in that it could be left to the mercy of hyaenas, elephants and fires. A total of 141 000 pictures (at 1-min intervals) of ostrich nests was taken. The photographs were analysed using

Plate 2.3 A male ostrich on his nest, being monitored by a time-lapse camera from a distance of *c.* 15 m. The grass cover conceals most of the bird from all but a very close observer.

a film editor with frame counter, enabling extensive information to be gathered on nest attendance by different birds, on nest predation and on egg movements.

c) *Visiting nests.* The nests were observed from a distance using binoculars to determine whether a bird was present or not (since an undisturbed bird on a distant nest keeps its head up) and then approached circumspectly and slowly in the vehicle. Any bird generally departed when the car was within 30 m. Usually it walked out of sight, but one or two birds became sufficiently accustomed to our approaches to remain within 200 m. The vehicle was juxtaposed between the bird and the camera, which was checked and the film or batteries changed as necessary. The vehicle was then moved to provide concealment while the eggs in the nest were examined (Section 2.4).

Visits to nests were kept as brief, infrequent and undisturbing as possible, consistent with obtaining the data needed. In general, nests where laying was taking place were visited every 2 days to examine the eggs; visits at 4-day intervals were required for keeping the time-lapse cameras running.

Plate 2.4   The author scratches a small number on the end of each egg.

d) *Studying nest predation*. The risks of destruction of nests by predators were investigated essentially in two ways. The period of time that each active natural nest survived was recorded and analysed (Section 5.3) and artificial dummy nests were set up and left unattended at appropriate places in the landscape, and their survival monitored and analysed (Section 5.3).

2.4 The Study of Eggs

a) *Marking eggs*. Each egg was individually numbered by scratching the shell at one end with a sharp spike. This indelible mark was inconspicuous yet easily found, and provided reliable identification, even after intervals of weeks or months. Within a nest, each egg was given a number from 1 onwards roughly in the order in which they were laid.

Two simple methods were also used as short-term labelling systems, in connection with describing and photographing eggs or nests. For convenience, when each individual egg was to be photographed, its number was written large and distinctly on the side using a soft lead pencil; this number was quickly and easily wiped off when the egg was returned to the nest. When the positions of the eggs in the nest were being recorded photographically, round metal discs – each clearly marked with a large number – were placed temporarily on the top of each egg to enable identification on film.

b) *Describing, measuring and recording eggs.* The following details were recorded concerning each egg examined in the nests:

- Nest and egg-numbers.
- Weight – measured by means of a set of kitchen-type scales.
- Length and width – measured using metal calipers.
- Long and short circumferences – measured using a metal tape-measure.
- Description of the surface – including such features as the shape, size and density of the shell pores and the extent to which they were found at the ends of the egg; surface colour of the egg; general shape of the egg including particular abnormalities; and any clear similarities to other eggs in the nest.

Most of these features were also recorded on film, by photographing each egg on a specially constructed calibrated support, so that lengths and curvatures could subsequently be determined from photographs, and a leisurely examination of pore appearance made.

Repeat weighings of individual eggs enabled data to be collected on the rate of weight loss under different conditions.

c) *Determining motherhood of eggs.* Observations were made of birds laying to establish which female was the parent of which eggs in the nests under close study. A nest would be watched discreetly from a considerable distance until a bird was observed to lay there. Laying (described in Section 4.2) was distinctive to the practised observer. As soon as the laying bird had departed, we would approach and identify her and then go to the nest and mark the newly laid egg; it was easily recognizable for some time by being clean, creamy coloured and warm.

This procedure was elaborate and time-consuming, and risked disturbing the laying birds. It eventually proved necessary to carry it out only once per female per nest. We discovered that individual female ostriches were relatively consistent in the type of egg they laid (Section 6.2) – all those laid by a particular female would be similar in the details of their surface appearance and in their general size and shape. A similar egg would appear in the nest regularly on alternate days, corresponding with the days on which that particular individual visited and laid in the nest. With experience, even if a nest was found only fairly late in the laying period, it proved possible by observation to identify which individual

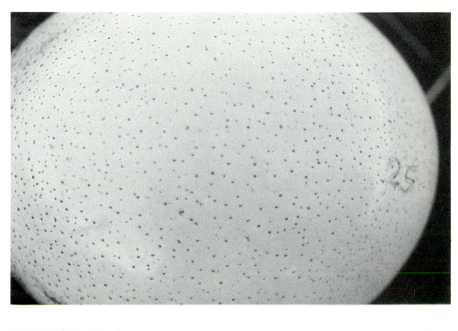

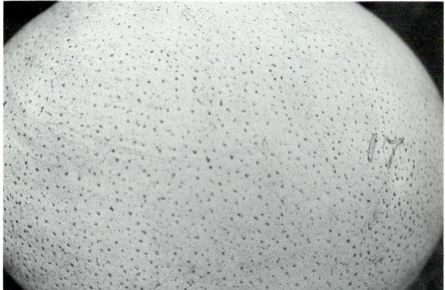

Plate 2.5 Eggs laid by two different females in the same nest. Slight differences in the smoothness and pore patterns can be seen.

coming to it was responsible for which egg of each of the most common types. Then, by comparison between eggs, it was possible to group together with a considerable degree of confidence all those which had been laid by the same individual.

In some cases, there were very clear differences between individual females in the types of eggs they laid, and in some cases rather little. We were perhaps fortunate in the fact that some of the most closely studied birds laid nicely distinctive eggs. But it should be emphasized that the differences involved were in fact only quite minor variations among basically large white, round eggs.

b) *Recording the movement of eggs in the nest.* At each visit to the nest, a rough sketch was made of the position of the eggs relative to one another. Greater accuracy was also achieved by photographing the nest from above, after temporarily marking each egg with a numbered metal disc (see above). When an egg was removed from the nest for weighing or measuring, its disc was left there to mark the exact position to which it was to be returned.

c) *Measuring egg temperatures.* The temperatures of unattended eggs were measured during experiments described in Section 5.4 using simple thermometers, which were inserted through holes drilled into the shell.

The temperatures in active nests proved considerably more difficult to measure. The birds tended to peck at wires attached to the eggs, and to move them around within their nests. Two methods were used, each with limited success. At two ostrich farms at Oudtshoorn, thermistor probes were inserted into a fibreglass egg which was tethered by a short wire in the centre of a group of eggs being incubated. Some birds pecked and scratched at this fixed egg such that records could only be made for about 24 h at a stretch. In one year, in Tsavo, an automatic temperature recording device (designed and produced by H & B Developments of Cambridge) fitted inside a fibreglass shell was used to record temperatures around the time that incubation commenced. This battery-operated device recorded temperatures electrically at 30-min intervals on a roll of heat-sensitive paper and could be left to run for periods of up to 5 days. Although realistic in appearance, this artificial egg was one of the first to be discarded by the ostriches when incubation began, so few data were obtained from it.

3 The Population

Fighting

In order to investigate the relative breeding success of different ostriches, it was necessary to determine the categories of birds that were distinguishable and the proportions of the population falling into those age and sex categories. The relative advantages of different reproductive strategies depend partly on the sex ratio of breeding birds within the population; biases due to differential conspicuousness and distribution make the sex ratio much more difficult to determine than might at first be imagined with such large birds.

Breeding success depends also on the male and female being together for at least part of the time, and therefore an examination of ostrich grouping patterns was undertaken. Getting together with known individuals is more straightforward if those individuals restrict their movements within defined geographical ranges. The ranges of males and females have different functions and might therefore be expected to have different shapes according to their purpose; an examination was carried out of the distribution of sightings of known individuals.

3.1 Age Structure

All ostriches sighted during the study periods were classified into one of the three age categories described in Section 4.6, namely adults, immatures (2 years old) and juveniles (1 year old). Chicks (less than 3 months old during the study periods) are ignored for present purposes.

Table 3.1

Age structure of the population

	Year			Mean % 1977–79
	1977	1978	1979	
Total number of birds sighted	1535	981	849	
Number of adults (3 or more years old)	892 (58.1%)	690 (70.3%)	720 (84.8%)	71.1
Number of immatures (2 years old)	5 (0.3%)	249 (25.4%)	37 (4.4%)	10.0
Number of juveniles (1 year old)	638 (41.6%)	42 (4.3%)	92 (10.8%)	18.9

Table 3.1 shows the proportions of these three categories among all ostrich sightings in each of the three years. It should be realized that the actual number of individual birds involved was small, and that most of the sightings were therefore repeat sightings of the same individuals (but resightings within 4 h are excluded). Unlike the other categories, juveniles moved around as large conspicuous compact groups of up to 25 individuals, which split and rejoined occasionally. These juvenile groups moved fast, and moved over a larger area than that used by either adult males or females. I do not place great reliance on the figures in Table 3.1, but present them as a rough indication only of the age structure of the ostrich population in Tsavo West National Park.

The following points can be made:

1. The overall average proportions of the population were 71% adults, 10% immatures and 19% juveniles.
2. These proportions fluctuated considerably from year to year. This fluctuation may have been caused partly by the erratic and unpredictable movements of the juvenile groups into and out of the intensive study area; in addition, however, differential nesting success between years probably contributed to the change in juvenile numbers. The juvenile proportions are almost certainly too high, partly because their groups were conspicuous and partly because they tended to frequent one particular area where for other reasons they were disproportionately likely to be observed.
3. The high proportion of juveniles in 1977 was followed by a high proportion of immatures in 1978, which reinforces the likelihood that the

Plate 3.1   A group of juveniles, about 1 year old, with Mt Kilimanjaro behind.

juvenile percentage of the population in 1977 provided an indication of the previous year's nesting success.

4. However, the immature proportions in 1978 were lower than the juvenile proportions in the preceding year. While this might be an indication of mortality or expulsion between 1 and 2 years of age, it probably represents also a bias due to the difficulty of reliably distinguishing immature females from adult females. Of the 189 sexed immatures, 75% were recorded as being male; it is likely that recording some of the immature females as adult females contributed to a greater skewing of the observed adult sex ratio in 1978 than in 1979 (see below).

3.2 Sex Composition

Determination of the sex ratio among adult birds is an essential prerequisite for understanding the communal nesting system. However, there are considerable problems involved in determining the sex ratio.

There was clearly a skewed sex ratio. Table 3.2 shows for each year the numbers of adult birds of each sex sighted. Again, resightings of the same individuals within 4 h are excluded. Birds which were seen at known nests

Table 3.2

Sex ratio of adult ostriches computed by different methods

	Year			
	1977	1978	1979	Mean 1977–79
1. Total numbers of birds sighted				
Males	373	297	353	
Females	519	393	367	
Sex ratio[a]	1.39	1.32	1.04	1.25
2. Numbers of birds detected within 375 m				
Males	60	32	74	
Females	98	55	101	
Sex ratio[a]	1.63	1.72	1.36	1.53
3. Numbers of birds within 400 m of the vehicle's route				
Males			74	
Females			94	
Sex ratio[a]			1.27	
4. Numbers of birds within 200 m of the vehicle's route				
Males			49	
Females			70	
Sex ratio[a]			1.43	
5. Numbers of individually recognized birds within the study area				
Males			14	
Females			24	
Sex ratio[a]			1.71	

[a] Number of females per male.

were also excluded. From this first analysis, a slightly skewed sex ratio of 1 male to 1.25 females is produced.

There is a major source of bias in this result, due to the fact that the males with their striking black (and partly white) plumage are much more conspicuous than the cryptically coloured females. Thus a higher proportion of the males present would be expected to be detected and therefore counted. I examine the extent of this bias, and then attempt to correct for it.

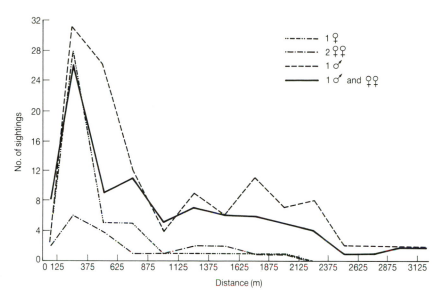

Figure 3.1   The distribution of the distances at which ostrich groups of different compositions were first detected.

For each sighting of each group, we recorded where practicable which individual in the group was sighted first. This was possible with 256 mixed groups (i.e. those containing a bird or birds of both sexes). The male was detected first in 78.5% of the observations, despite the fact that males made up only 40.2% of the individuals in the 477 mixed groups encountered.

Further indications of the greater conspicuousness of males is given in Fig 3.1 and Table 3.3. Figure 3.1 shows the distribution of the distances at which birds were detected in 1979. It is apparent that a considerably greater proportion of males than females was first detected at distances greater than 1 km. Few birds were first detected when closer than 100 m, mainly because they would usually have been detected at an earlier stage during the approach of either the observer or the ostrich.

The great variability in the ranges at which birds were detected was caused not only by the differing degrees of conspicuousness, but also by different methods of detection (such as binoculars from a stationary car, as compared with the naked eye while driving), different amounts of cover in different places, and features such as hills, which obviously made it impossible to see a bird until it was close at hand. On average, a single male was usually detected almost twice as far away as a single female. An

Table 3.3

Mean distance of detection of adult ostrich groups of different composition (number of groups in parentheses)

Composition (and no. of groups)	Mean distance (m)			Mean 1977–79
	1977	1978	1979	
1 male	940 (96)	789 (76)	1022 (125)	936 (297)
1 female	546 (51)	546 (47)	454 (46)	517 (144)
2 or more females	795 (34)	686 (18)	734 (22)	750 (74)
1 male and 1 female	752 (68)	846 (56)	882 (55)	821 (179)
1 male and 2 or more females	934 (38)	840 (25)	1053 (34)	951 (97)

increase in the number of females in a group, and the addition of a male to it, both increased the range at which it was first detected (Table 3.3).

Thus males were indeed more conspicuous than females, as indicated by the facts that they were usually detected when further away from the observer, and that they were usually detected before the females they were accompanying. It is clear, even if at first sight surprising with such large birds, that many potentially detectable ostriches were not detected by the observers.

In determining the adult sex ratio, I have attempted to overcome the conspicuousness bias in two ways. One method was to consider the sex ratio only among those adult birds which were first detected when within 375 m. This distance was chosen because Fig 3.1 shows that 67% of single females, the least conspicuous category, were first seen only when within this range, and there was a sharp drop-off in the numbers of females seen at greater ranges. As Table 3.2 also shows, this method produced a sex ratio ranging between 1 male to 1.36 females and 1 male to 1.72 females. The assumption here is that most birds were located and were detected to the side of the vehicle's route, rather than being detected in front. If a significant proportion were in fact in front of us and could have been approached much closer, to disregard birds which were detected more than 375 m ahead would have biased the data in favour of the less easily detected females.

The second method was to consider only those birds which were within a strip of fixed width on each side of the track along which the observer was travelling, regardless of the distance away at which they were sighted. The necessary observations were made only in 1979, and are summarised in Table 3.2. Decreasing the strip width from 400 m to 200 m resulted in an increasingly skewed sex ratio. Since there was no reason to assume that females tended to stand closer to the route than males, this suggests that 400 m was an excessively broad strip, and that some females were being missed. Using a strip 200 m wide on each side of the track produced an adult sex ratio of 1 male to 1.43 females. Even at such close range, however, one cannot by any means guarantee to see every bird, even if they are standing, because they are often completely concealed by bushes.

All these methods have another inherent bias in that because the study took place during the breeding season, at any time some individuals were on their nests and therefore could not be included in the count. Birds sighted on known nests were not included, because the frequency of sightings of those individuals would have depended only on the frequency of our visits to those nests. However, their mates were counted if they were seen elsewhere. Our observations tended to take place in the afternoons more than the mornings, and thus were possibly a little more likely to be made at times when the female rather than the male was on the nest (see Section 4.2). However, the bias is small, partly due to the fact that only a proportion of birds had a nest at any one time. To have excluded both birds of a pair known to have a nest would have produced another bias in magnifying whatever skewed sex ratio there was.

Another method of determining the sex composition of the population was simply to count the number of individuals recognized within the study area. Again this was possible only with the 1979 data, although the information from previous years supports these data. The totals were 14 identifiable males and 24 females, giving a sex ratio of 1 male to 1.71 females. Conflicting biases arise here for two reasons. First, a proportion of birds seen within the study area were not identified for a variety of reasons, but usually because they could not be approached close enough in the time available. Males were generally more easily recognized than females, as described in Section 2.2. Second, and of a more serious nature, there are significant edge effects, because females move over a considerably larger area than males because they are not confined to territories (Section 3.4). Thus even if the sex ratio were 1:1, a greater number of different individual females than males would be expected to enter the study area.

I conclude that the average adult sex ratio in the population was of the order of 1 male to 1.4 females, and I have used this figure in subsequent calculations. The causes and the effects of this skewed adult sex ratio are discussed in Section 9.5. Here it is worth clearing up one small point. As described above, 2-year-old females (immatures) could not be reliably distinguished from older females, and it appears that a number of them may have been counted as adult females and thus have contributed to the skewed sex ratio observed. However, some females are able to breed at 2 years old, so it is not unreasonable for them to be taken into account in assessing the breeding sex ratio.

3.3 Grouping

All birds sighted during the study periods were classified as to whether or not they were part of a group when first seen. A bird within 150 m of another was allocated to the same group; two birds further apart were assigned to different groups unless the presence of other birds reduced the gaps between birds to less than 150 m.

Ostriches tend most often to be solitary birds. Figure 3.2 shows the overall distribution of group sizes of adult ostriches for each of the three years; most sightings (51%) were of single birds. If we assume that the time of first sighting was independent of group formation or break-up, we can calculate the time birds spent in groups of different sizes. On average, a bird would spend about 29% of its time alone, 37% with one companion, 19% with two companions and 15% with three or more. Few groups (5%) contained more than one male (Fig. 3.2). Birds in the company of others were usually females, who were slightly more likely than not to be accompanied by a male.

Table 3.4 shows the group sizes and composition in detail: they were similar in each year. Groups were loose in the sense that they changed in size quite frequently, by the arrival or departure of one or more birds. There were 66 changes in group size during 139.5 h of continuously watching females, indicating a change on average every 2 h. A few male – female pairings, or of two females, lasted for several days, but most groupings were much shorter lived.

Thus in general adult ostriches during the breeding season in Tsavo West National Park may be considered to be fairly solitary birds, which none the less do spend time with a few temporary companions. Ostrich groups were also loose in the sense that the individuals in the group were usually quite widely dispersed, being rarely within 5 m of each other and more usually 10–30 m apart. It must be stressed that the above refers only to adult birds during their breeding season when the observations were made. Outside the breeding season, ostriches are known to be more gregarious. The juvenile birds were also much more gregarious. They

Table 3.4

Composition of all adult ostrich groups

	1 bird		2 birds			3 birds				4 birds				5 birds				6 birds	7 birds
				1M			2M	1M		3M	2M	1M		3M	2M	1M			
Year	1M	1F	2M	1F	2F	3M	1F	2F	3F	1F	2F	3F	4F	2F	3F	4F	5F		
1977	162	95	8	106	54	2	3	43	9	0	2	17	0	0	2	7	0	0	0
1978	106	66	13	96	27	0	4	28	6	1	1	15	1	1	0	4	1	3	2
1979	172	75	12	82	28	3	3	40	4	1	0	11	1	0	2	3	0	0	0
Totals 1321	439	236	33	284	109	5	10	111	19	2	3	43	2	1	4	14	1	3	2

Numbers of groups with

M: male; F: female.

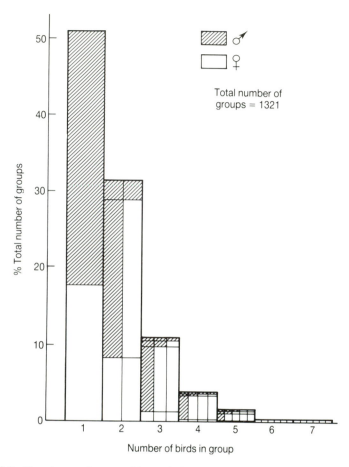

Figure 3.2   The sizes and compositions of all adult ostrich groups sighted.

formed much larger groups, within which they were much closer together, such that 20 birds or more might all be within as many metres of one another.

3.4 Distribution

It quickly became apparent that during their breeding season, individual male ostriches were to be found resident in the same area for a period of at least 3 months. Figure 3.3 shows the points at which nine male ostriches were observed during 1979, in order to give a general picture of the size of the territory inhabited by each male. Some 64% of the points indicate

Plate 3.2   One of the largest groupings of adults encountered — a male and four females.

the locations where the birds were first sighted on any day. During long periods of observation of individual males, some birds travelled considerable distances; the remaining 36% of points on the map were repeat observations of the same bird on the same day, at intervals not less than 30 min apart – they have been included only when the bird's location changed appreciably during the observation period and when the new location resulted in extra information on the area used by that bird. The frequent crossings of the centre of males' territories have not been shown, nor have the numerous sightings of birds at their own nests. Therefore, the map should not be used to indicate the birds' intensity of use of different places within these territories.

Joining the outermost points where each male was regularly seen and from where he was not seen being expelled, gives territories ranging from 12 to 19 km^2 for six breeding adult males (Table 3.5). One non-breeding male (Stalin) successfully held a territory that was minute compared with those of the breeding males. Stalin did not have a nest, and was not observed mating despite his courtship displays. For two further breeding males (Grit and Moth) at the edge of the intensive study area, the full extent of the territory was not known.

There was relatively little overlap in the areas used by the different males

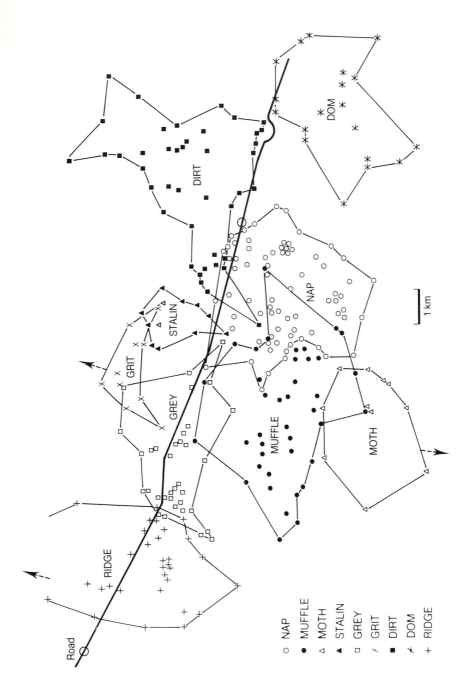

Figure 3.3 Territories of male ostriches. The map shows the locations where nine male ostriches were observed during 1979. The outermost points for each male have been connected.

Table 3.5

The territory sizes of nine adult
male ostriches in 1979

Male	Territory (km^2)
Dirt	19.1
Dom	17.3
Grey	11.9
Grit	>6.3
Moth	>11.5
Muffle	14.3
Nap	18.1
Ridge	16.1
Stalin	3.5

Mean (excluding Stalin, Grit
and Moth) = 16.1 km^2.

(see Fig. 3.3) and these areas may therefore reasonably be termed territories. It was clear that aggressive interactions between adjacent males were the cause of their remaining within their territories but often also the cause of their excursions. Territorial interactions were particularly frequent during the early part of the breeding season, when birds intruded more into their neighbours' territories. A male, on seeing another male anywhere within 1 km of it, usually strutted or trotted towards its adversary, this sometimes taking it outside its own territory. Fights were rarely seen, but prolonged chases were common, and again often resulted in the chaser ending up well outside its own territory, especially if it was pursuing an immature male who did not attempt to dispute with the owner. The chaser would then in turn be driven back by the territorial owner. Oscillations across boundaries due to intrusions were seen on a large scale, and the boundaries were clearly not marked or fixed. At a balance point, at any particular time, the rival males spent up to 2–3 h within about 150 m of one another, exhibiting a great deal of wing-flagging and parallel walking.

Males tended to occupy the same areas in successive years. Only one (Muffle) of the eight breeding males intensively studied in 1979 had shifted detectably from the area used in the two preceding years; he had moved eastwards and vacated the western third of his territory in favour of male Grey; otherwise, he too was occupying his usual territory.

Making study visits only during the breeding seasons, we were unable to collect data on land tenure outside the breeding season. It is likely that the territorial system breaks down outside the breeding season, but it is not

Plate 3.3   A breeding male's size, stance and black plumage make him conspicuous in his territory.

known where the males then go. It is known that they lose the bright red flush on their neck skin, and it is known that ostriches escorting chicks travel far outside their breeding territories (Hurxthal, 1979). During late October at the end of the 1978 study visit, the territorial system appeared to be disintegrating – males were losing colour, they were rarely mating, and they were intruding and tolerating intrusions into territories to a much greater extent than two months previously. Thus the fact that males held the same breeding territory in successive years was apparently not due merely to their holding it throughout the intervening period – they probably had to re-establish their territorial claims.

Females travelled over considerably larger areas than males. Figure 3.4 shows the locations of five frequently observed adult females, all of whom were known to have laid during 1979. Their ranges averaged 25.8 km² (Table 3.6), which was at least 1.5 times the size of the average breeding male's range. The females probably had still larger ranges in fact, because it is likely that they all travelled outside the intensive study area during those long periods when they were not being observed. Two females were indeed encountered outside the study area, despite the fact that the pro- bability of encountering them there was relatively low.

Comparison of Figs 3.3 and 3.4 shows that the extent of females'

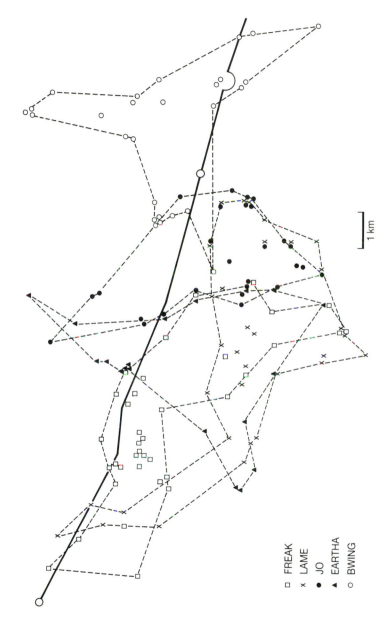

FREAK
LAME
JO
EARTHA
BWING

Figure 3.4   The areas used by adult female ostriches. The map shows the locations where five adult female ostriches were observed in 1979.

Table 3.6

The ranges used by five adult females in 1979

Female	Range (km^2)	Dirt	Dom	Grey	Grit	Moth	Muffle	Nap	Ridge	Stalin
				Male territories used, in whole or in part						
Bwing	24.0	•	•						•	
Eartha	25.1			•	•		•	•		
Freak	34.1			•		•	•		•	
Jo	14.5				•			•		
Lame	31.4					•	•	•	•	
Mean	25.8									

movements bore little relation to the males' territories. Females' ranges did not simply encompass the whole of the ranges of two or more males. On the other hand, some females did appear to avoid the territories of certain males. For example, female Lame was never seen in male Grey's territory, although her range stretched along both sides of it; it is not known whether she made a detour around it or hastened straight through it. In addition to those males whose territories they entered, females also met and interacted with other males near the latter's territorial boundaries.

The females' ranges in Fig. 3.4 bore little relation to one another – there was a great deal of overlap of the ranges of individuals occupying the same part of the study area. Different females used the territories of different combinations of males (Table 3.6). None wandered throughout the study area, but each was apparently confined to a large irregularly shaped part of it. Thus females during the breeding season made use of discrete, undefended, overlapping ranges, much larger than, and independent of, the territories of the males.

It is very striking that the shapes of the ranges of males and females were as different as their sizes. Whereas the males' territories, as befits defended areas, tended to be compact and with relatively short defendable boundaries, the females' ranges were more straggly and elongated, appropriate for encountering a variety of different males. The females' ranges, like the males' territories, tended to remain similar in successive years; this was less obvious than with the males, partly because the ranges extended outside the intensive study area, and partly because few females could be reliably recognized in successive years. Nevertheless, female Bwing, for example, was observed in only the eastern half of the study area in all three years, and female Jo only in the central part.

4 The Breeding System

Kantling

Having dealt with the adult population in Chapter 3, I now present general observations and data on their nesting. The intention at this stage is to provide a background of information against which later to carry out more detailed analyses of the reproductive options open to different categories of breeding birds. This chapter summarizes when and where ostrich nests in general were started, and by whom; how they grew; how they were attended; what other birds laid in them and when; how large they finally ended up; and how many of those eggs were incubated and by whom. It was very clear that there was a great deal of variation and that no nest could be considered typical. None the less, an attempt is made to quantify what is likely to happen on average in ostrich nests in Tsavo West National Park.

4.1 Before Nesting

As outlined in Section 3.4, males held and defended large territories during the breeding season. They sat inactive at night, but spent most of the daylight hours on their feet. Figure 4.1 shows the distribution throughout the day of various activities of pre-nesting male ostriches: the birds did not exhibit any particularly marked daily pattern. Nor did the females. On average, both sexes spent some 33% of their daylight hours feeding, 29% travelling, 18% being alert and 9% preening. Such behaviour was not consistent from day to day, nor from bird to bird. A male approached any

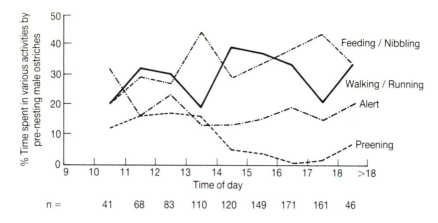

Figure 4.1 The distribution thoughout the day of activities by pre-nesting male ostriches.

adult ostrich he saw within his territory, usually driving it out if it was male and displaying to it if it was female. When both sexes were present among the newcomers, the territorial male's attention and displays were almost always devoted first and primarily towards the male among them.

Courtship behaviour by males towards females was performed particularly towards new arrivals. Indeed, a male which had been with a female for several hours was much more likely to display to her after they had drifted some way apart. As described in Section 1.6, the main courtship display consisted of 'kantling' – the male rocking from side to side, waving his wings alternately above his back. Whenever possible, we counted the number of waves in the often long-lasting and clearly energetic performance. Figure 4.2 shows the number of waves made in different circumstances: the males averaged 36 waves when courting females ($N = 84$), 39 waves before successful copulations and 29 waves in the 32% of cases where kantling was not followed by copulation (the difference is not significant, $P = 0.09$). The main cause of unsuccessful attempts at copulation (in 18 out of 21 cases) was that the female refused to squat; sometimes she merely stood, sometimes she ran away from the male. It was not possible to predict whether a kantle display would be successful or not in inducing a female to copulate.

Some 29% of matings were not preceded by a kantle display. The male simply approached the female in an immediately pre-copulatory posture with both wings held high above his back, and the female squatted at once and enabled copulation to occur. In a few cases, in the absence of the kantle display, a long chase ensued before the female was induced to

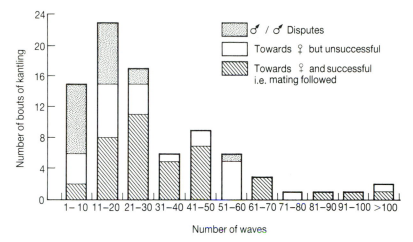

Figure 4.2 The distribution of the numbers of waves made by male ostriches on 84 occasions in 1979 when performing the kantle display to males and to females. Some kantling bouts aimed at females were successful in that they were followed by copulation, whereas the remainder were unsuccessful.

copulate. In several cases, the male boomed during his initial approach towards the female.

The kantle display was also used during agonistic interactions between males (Fig. 4.2). In these cases, there were fewer waves than when courting females (means of 13 *vs* 36 waves, $P < 0.001$). The male giving the display was almost invariably the one who at the time was driving the other male away, and the latter almost always turned and approached the displaying bird. The driving continued as soon as the display was over.

Courtship displays and matings were particularly likely to occur near scrapes. Males made several scrapes, widely scattered within the territory. They sometimes made them when alone and later showed them to females, and sometimes started making them when a female was with them. The process involved scratching with their powerful toes, and at intervals sitting at the site and apparently grinding the underside of their body (especially the callosities) into the soil. Mating often occurred just before or just after the scrape-showing. Within a period of 2–3 h, a male might show as many different scrapes to the same female, and over a period of days he showed them to a number of different females. Only a small proportion of scrapes were actually used as nests – 5 out of the total of 25 we observed.

Plate 4.1   A new scrape which may be selected to become a nest; scrapes are made with the powerful toe.

4.2 Laying

Eventually, one of the male's scrapes is accepted by a female and therefore becomes a nest when she lays the first egg in it. We established that in every case out of 27, the first female to lay in a nest was the one who subsequently undertook the guarding and incubating of it. She was termed the 'major' hen (after Sauer and Sauer, 1966a; Hurxthal, 1979). During the preceding few days at least, and sometimes considerably longer, a pair bond could usually be observed developing between this female and the male – they were usually to be found moving through the male's territory together.

The location of the nest was impossible to predict. It was not necessarily near the centre of the male's territory, and indeed was sometimes very close to the edge. The location bore no relation, positive or negative, to that of the previous year's nest, but none was within 300 m of it.

Almost without exception, female ostriches laid at regular 2-day intervals. Only four instances were detected of birds laying at 3-day intervals. Laying took place during the afternoon, the mean laying time being 15:47 h (range 14:16–17:48 h for 27 layings observed and timed in 1979). A bird that was about to lay stood at the nest for a few minutes,

Plate 4.2 A male leads a female, walking steadily towards what may turn out to be a nest.

its head alternately up and down among the eggs. The tail was raised during the 15 s or so before laying, and the bird then sat abruptly. The egg was laid during the 1 min that the bird remained sitting (Fig. 6.1). Having completed laying, the female would stand once more, her head among the eggs for a few minutes. During this time, from a hide near the nest, eggs could be heard as they clonked against one another while being moved about in the nest.

At the earliest stages, the nest was left unattended for a large part of the time, this proportion diminishing as the nest grew in size and age. Figure 4.3 shows (from time-lapse photography over 109 nest days) the changes in average nest attendance by the male and major female during the laying period. In general, a regular pattern developed gradually over three weeks, whereby finally the nest was constantly attended by the female from about 09:30 until 16:30, and by the male for the rest of the time. During this period of increasing attendance, the nest was more closely attended in the middle of the day than during the early morning or late afternoon. The female's attendance pattern apparently developed a little sooner than the male's. There was, of course, considerable variation among individual nests and among birds. The standard deviations in attendance at those six

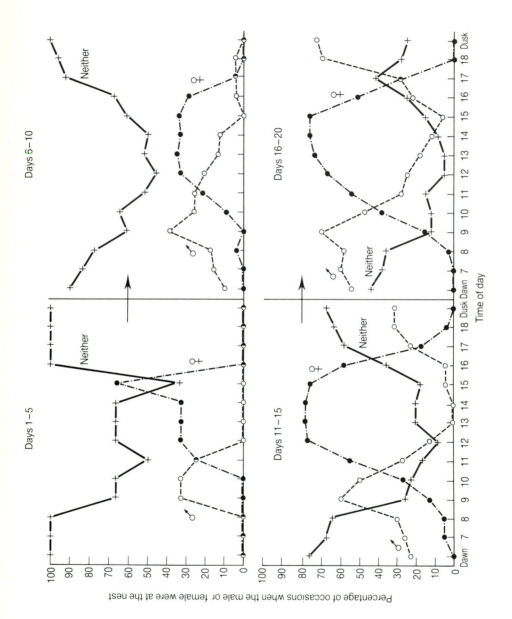

Days 6–10

Days 1–5

Days 16–20

Days 11–15

Neither

Neither

Neither

Neither

♀♂

♀♂

♀♂

♀♂

Percentage of occasions when the male or female were at the nest

Time of day

Dawn 7 8 9 10 11 12 13 14 15 16 17 18 Dusk

0 10 20 30 40 50 60 70 80 90 100

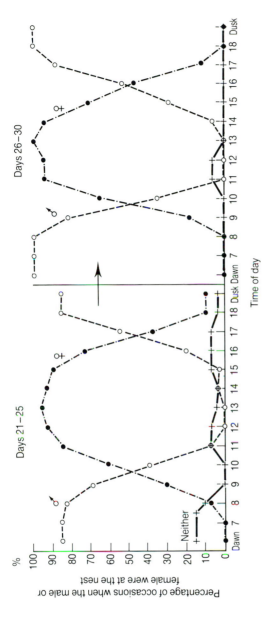

Figure 4.3 The development of the pattern of attendance at the nest, with increasing age of the nest. The data come from 109 nest-days of records at 17 different nests. The graphs show, at dawn, dusk and on the hour at each hour, the percentage of cases when the male, the female or neither bird was at the nest.

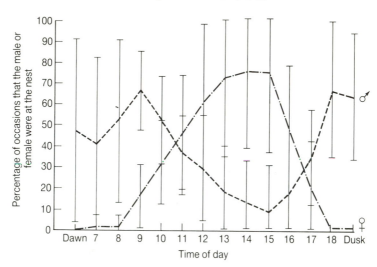

Figure 4.4 The variation in attendance pattern at nests during the second half of the laying period. Means and standard deviations are plotted for attendance by the male and female on the hour at each hour. The data come from 49 nest-days at 13 different nests.

nests where the samples were large enough (> 4 days at each nest) at the most variable period (days 11–20) of nest growth are shown in Fig. 4.4; even at this time, the differences in the probability of the male's or the females's attendance at the nest at most times of the day were statistically significant.

A bird attending a nest on its own sat on the nest for the bulk of the time, for periods usually between 15 and 90 min. It would then stand for a few minutes and turn itself or its eggs. The eggs were below the temperature necessary for incubation at this stage, not rising above 32°C even during the hottest part of the day. The bird's presence probably helped to protect the eggs from both heat and predators (see Chapter 5).

A difference in the attendance pattern could be seen, particularly during the early stage of the laying period, between those days on which the major female laid and her 'off-days'. As Fig. 4.5a shows, the female was considerably more likely to be at the nest during the middle of those days on which she laid than on those days when she did not. The increase in probability of her attendance started to show itself at least 5 h before she was due to lay. On the days on which she did not lay, the male was more likely to be there instead. It is not clear whether he was there mainly because the female had not turned up to relieve him or whether he too exhibited his own 2-day rhythm; there are indications that the latter was the case. There were still signs of this phenomenon when the nest was almost full (Fig. 4.5b).

During the early laying stages, the male and the female were quite often at the nest together; later they rarely were, but simply swapped places rapidly. Figure 4.6 shows the probability of attendance at the nest by both sexes related to the age of the nest. This was measured as the percentage of the daily total number of 1-min exposures on film in which both birds were visible. Days on which fewer than 360 min of film was recorded were discarded from the analysis. The values shown in Fig. 4.6 are the averages for a number of different days at 3–10 different nests. Figure 4.6 also indicates what the birds were doing when both were at the nest. Soliciting by quivering the lowered wings, thereby drawing attention to the nest site, was prevalent early on, but declined with increasing age of the nest, as did the probability of a kantle display or mating near the nest.

It should be borne in mind that the data were obtained from time-lapse photography using cameras usually set up 10–20 m from, and pointing at, the nest. Birds other than those being recorded on film may have been nearby and influenced the behaviour of those birds we were filming. Also, by using time-lapse photography, it was not always possible to distinguish between the major hen and other hens who visited her nest. However, these visits were brief and any errors due to them scarcely influence the data in Figs 4.3–4.6.

4.3 Minor Females

Not only were female ostriches seen to visit other females' nests, they often laid eggs in them as well. A bird which laid in another female's nest was defined as a 'minor' hen (after Sauer and Sauer, 1966a; Hurxthal, 1979). Who these minor females were is considered in Section 7.1. They were termed 'minor' females because of the minor role they played at the nest – they spent 15–30 min at or near it on alternate days, they laid eggs in it, but they took no part in attending or guarding it, nor later in incubating it. The major hen was reliably distinguished from the minor hens by the facts that she laid the first eggs, that only she sat for any length of time at the nest (other than for a couple of minutes when laying), and that only she attended the nest without laying in it on that day. The minor hens were tolerated at the nest for reasons considered in Section 6.1. When a minor female arrived, the major female usually got up within a couple of minutes, and wait 5–20 m from the nest until the minor female had laid and left.

How the minor females usually found the nest was not clear. Some, certainly, were led to it by the male, with occasional intermittent displays

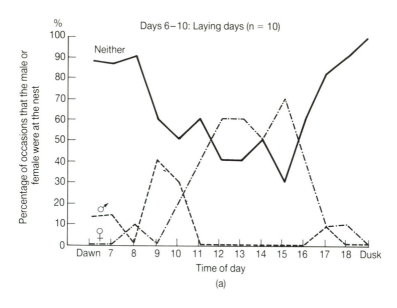

(a)

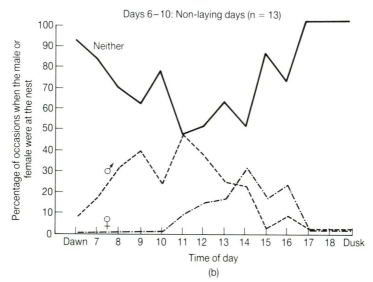

(b)

Figure 4.5   The attendance patterns at the nest of males and females on the days when the female did or did not lay.

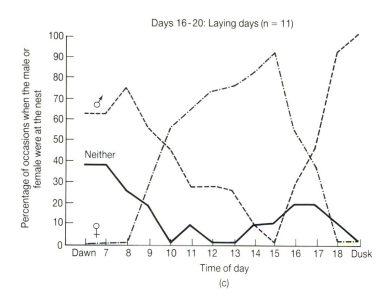

Days 16–20: Laying days (n = 11)

(c)

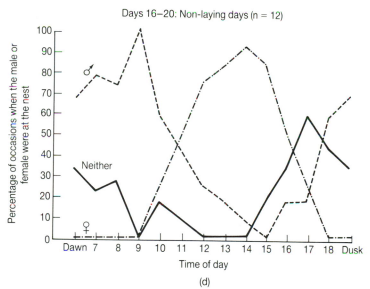

Days 16–20: Non-laying days (n = 12)

(d)

Figure 4.5   Continued.

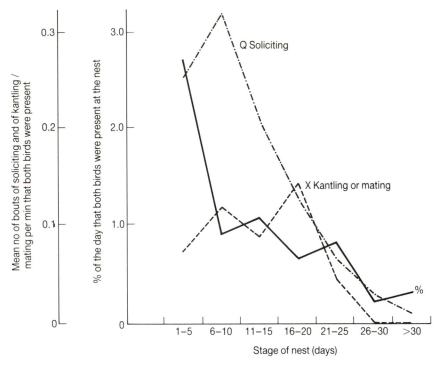

Figure 4.6 Probability of attendance at the nest of males and females together related to the age of the nest. The probabilities per minute of soliciting and of kantling or mating when both birds were present are also shown. The data come from 196 nest-days at 19 different nests.

of soliciting with lowered wings quivering when near the nest. It seems probable that in addition there were finer, more subtle signals by which a male indicated to a female that he was going to his nest. This was suggested by observations of a male walking steadily and apparently in a normal fashion towards his nest, followed by a female who had not laid in it but who subsequently did. However, I cannot be certain that this particular female had not been led to the nest before in a more obvious fashion. Minor females might also find a nest by following other females – including the major female – but unlike males, females did not exhibit any detectable nest-showing display to other females. Some of the soliciting displays in Fig. 4.6 occurred when males were showing a nest to minor females.

In only 14 nests, found early enough and surviving until incubation, was it possible to determine the number of minor females laying. Table 4.1

Plate 4.3 A male and three females at a nest in which all three are currently laying. From a portable hide only 30 m away, the eggs are not visible.

shows that the mean number of minor females was 3 +, with a range from 1 to 5 +.

On day 1 the major female laid her first egg. The minor females first began to lay between days 2 and 30, (with a median of day 5: Table 4.2). By day 5, the major female would have laid three eggs. A bias may have arisen due to the high rate of nest destructions: excluded from Table 4.2 are those nests which had not been laid in by minor females before they were destroyed, which biasses the data in favour of those nests in which minor females began to lay early. However, these accounted for only four of 29 nests, so they could scarcely affect the median date of first laying by minor females.

Nests varied considerably in their rate of growth, depending on how many minor females were laying, how many eggs each laid, and on how early they began to lay. It is useful to compile a picture of the rate of growth of an average nest. However, complications arise due to the piecemeal nature of some of the data – a few nests were discovered so late that their age could not be accurately determined, and a large proportion of the nests was destroyed in the middle of the laying period (see Section

Table 4.1

Numbers of minor females laying in different nest ($\bar{x}$ = 3+)

Year	Nest	No. of minor females laying
1973	A	4
1977	2	4
1978	1	2
	4	4
	7	4+
	15	5+
1979	1	3+
	2	3
	4	2
	6	1
	8	2
	10	2
	11	4
	13	2

5.3 for data on predation rates). None the less, Fig. 4.7 shows the mean number of eggs laid in a nest by the end of each day.

Clearly, minor females did not lay in nests at a steady rate, but laid faster in larger nests. The likely reasons for this are dealt with in Section 7.2. The slowing of the rate of appearance of eggs after about day 20 reflects the fact that at a number of nests incubation was beginning: this coincided with a fairly abrupt cessation of laying by both the major and minor hens.

For reasons that were not clear, most minor hens' eggs were laid on those days on which the major hen did not lay any eggs. In 20 nests, a total of 184 minor hens' eggs were laid – it was not known on which day 49 of these were laid. Of the remaining 135 eggs, only 31 (23%) were laid on the major hen's laying days in that nest, compared with 104 (77%) on her off-days. In 14 of the 20 nests, the number of eggs laid on the major hen's off-days was greater than on her laying days; in three nests, the reverse was the case, and in three nests the numbers were equal. Using a two-tailed Wilcoxon test, these differences were found to be significant ($P < 0.01$).

For nine nests that were known from an early stage and which survived until incubation, the duration of the laying period was determined: it averaged 23 days (i.e. incubation started on day 24), with a range of 18–33 days. In four of these nests, the last egg laid was laid by the major hen. In three nests, it was known that embryo development, and therefore

Table 4.2

The ages of nests when minor females first started to lay in them (median = 5 days, $\bar{x}$ = 7.3 days)

Year	Nest	Day first minor female's egg laid
1973	A	2
1977	2	3
	3	8
	4	4
	8	3
	9	7
	12	8
	15	4
1978	1	6
	2	8
	4	4
	5	2
	8	10
1979	2	3
	3	2
	4	23
	5	10
	6	30
	8	2
	9	3
	10	16
	11	5
	12	3
	13	8
	14	8

presumably incubation, had already started before the last egg was laid, and in three nests, 1–4 more eggs were laid in the nest much later, at least a week after incubation had commenced. Even without these extra additions, it is clear that there was a considerable range in the duration of the laying period (possible reasons for this are discussed in Section 8.4).

4.4 Nest Size

For only nine nests was it possible to determine the total numbers of eggs laid in them by major and minor females. The means were 11.3 (range 8–16) eggs laid by the major hens and 11.1 (range 3–20) by all the minor

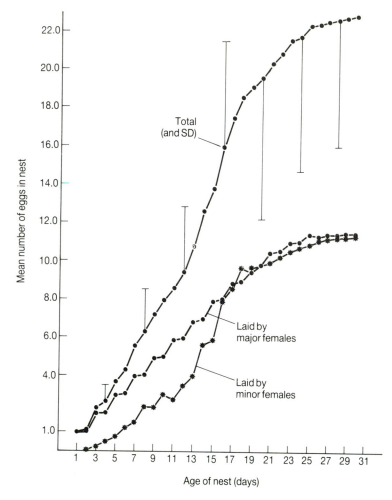

Figure 4.7 The mean rate of growth of ostrich nests. The total number of eggs is the sum of those laid by the major and minor females. The data come from 26 nests on day 1, decreasing to 10 nests by day 29. The large standard deviations shown are due in large part to the variation in the numbers of minor females laying in different nests.

hens. There was no significant correlation ($r = -0.16$, $P > 0.5$) between the numbers of eggs laid by major and minor females in any nest.

The figures quoted above refer to the total numbers of eggs laid in a nest, not to those remaining unbroken. Breakages of eggs occurred in nests at any stage. In 16 nests which reached the incubation stage and where it was known whether breakages had occurred during the laying stages, 8

(50%) suffered breakages; the mean number of eggs per nest broken during the laying period was 1.13.

Most of the information (five of the nine nests above) on the numbers of eggs laid by major and minor females was obtained during 1979 when nest sizes tended to be somewhat smaller than in the previous 2 years. Over the 3 years, the number of major hens' eggs per nest remained constant. The probable numbers of major and minor hens' eggs laid in all 20 incubating nests in all years are presented in Table 4.3. The figures in Table 4.3 refer to the number of whole eggs in the nests at the first observation after incubation had begun; in the case of nests which were only discovered after incubation had started, it was impossible to determine whether any breakages had occurred during or after the laying period. If one assumes that the nests for which data were not available each contained a number of whole major hens' eggs equal to the mean of the 9 nests above ($n = 11$), then the probable numbers of eggs laid by the minor hens can be obtained by subtraction, which gives an average of 14 eggs (range 3–25).

It was noticeable that at most incubating nests not all the eggs were actually incubated. A compact group of about 19 in the centre of the bowl were warm and covered by the sitting bird; the remainder were situated in a loose outer ring outside the nest bowl and 1.5–2.0 m from the central group. These eggs remained here, were not incubated, did not develop and eventually rotted or were destroyed. The nest had by this stage grown considerably from its original small shallow scrape – it was now an expanse of bare loose dusty soil some 3 m or more across, due mainly to the pecking of the major hen. Which eggs were moved to the outside ring, and by which bird, is considered in Section 6.2.

The number of eggs in the centre of an incubating nest was relatively consistent between nests, with a mean of 19 (range 14–25: see Table 4.3). By comparison, the number in the outer ring was more variable, ranging from 0 to 15 with an overall mean of 5.6 (or a mean of 6.9 in the 80% of nests which had eggs in the outer ring at all).

Thus we can summarize the history of an average surviving ostrich nest in Tsavo West National Park as follows. The major hen laid 11 eggs at 2-day intervals. Over this 3-week period, an average of three minor females also laid in it, producing about 14 eggs between them. At incubation, about 19 eggs remained in the centre and were incubated, while the remainder were pushed into the outer ring and not incubated.

Table 4.3

The sizes of incubating nests

		No. of eggs laid			No. of eggs		No. of eggs	
Year	Nest	Whole	Broken before incubation	Total	Centre	Outer ring	Major hens	Minor hens
1973	A	27	0	27	19	8	9	18
1977	2	20	0	20	20	0	11	9
	13	22	?	(22)	20	2	(11)	(11)
	14	36	0	36	22	14	(11)	(25)
	16	23	(1)	24	16	7	(11)	(12)
	17	36	(3)	39	21	15	(11)	(25)
1978	1	24	0	24	19	5	13	11
	4	29	2	31	18	11	10	19
	5	20	8	28	14	6	(4)	(16)
	6	34	0	34	21	13	(11)	(23)
	9	25	?	(25)	22	3	(11)	(14)
	11	20	2	22	18	2	(11)	(9)
	14	24	?	(24)	20	4	(11)	(13)
	15	31	1	32	18	13	(11)	(20)
1979	2	28	1	29	25	3	12	16
	4	15	0	15	15	0	11	4
	6	17	2	19	17	0	14	3
	8	16	0	16	15	1	8	8
	11	25	1	26	21	4	(11)	(14)
	13	21	1	22	21	0	10	11
All nests (n = 20)				(n = 17)				
Mean		24.7		26.1	19.1	5.6		(14.1)
(s.d.)		(6.1)		(6.7)	(2.7)	(5.0)		(6.1)
Early known nests (n = 9)								
Mean							10.9	11.0
(s.d.)							(1.8)	(5.4)

? in column 4 indicates that it was not known whether any eggs were broken.
( ) in column 4 indicates that eggs were broken but that it was not known whether the breakages occurred during the laying period or the incubation period.
( ) in columns 8 and 9 indicates estimated numbers (see text).

4.5 Incubation

The exact date at which incubation started was difficult to determine, particularly as we made only intermittent daytime visits to the nests. Often it had to be estimated, with an accuracy of only about 3–4 days.

Plate 4.4   A female incubates a large nest, with some of the surplus eggs visible in the unincubated outer ring around her.

Incubation was shared between the male and the major female with the male sitting on the nest overnight, and therefore for about 71% of each 24-hour period. The daily changeovers took place on average at 09:30 and 16:25 h. Figure 4.8 shows that the daily pattern during incubation was much the same as that during the late stages of laying (cf. Fig. 4.3), but with nest attendance almost equalling 100%. The changeovers were usually rapid, so that both birds were rarely present together at the nest and (as is shown in Fig. 4.6) displays rarely if ever occurred. By the middle of the incubation period, the male had lost the bright red neck flush distinctive of breeding males, and was rarely to be found interacting with any females.

The temperature of eggs being incubated was measured by means of a temperature recording device inside a fibreglass dummy egg (as described in Section 2.4). A series of recordings over a 96-h period early in incubation showed that the central temperature was being maintained at about 34–36°C. Other information regarding incubation collected from ostrich farms in South Africa, showed a slightly lower incubation temperature (mean 33°C) and nest humidity around 42% (Bertram and Burger, 1981a).

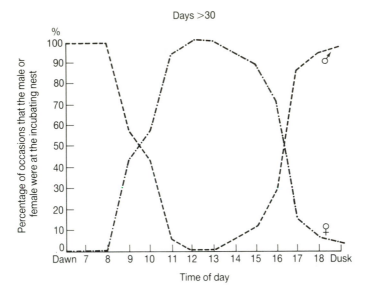

Figure 4.8 The nest attendance pattern of male and major female ostriches at incubating nests. Symbols as in Fig. 4.3. The data come from 35 nest-days at four different nests.

Incubation has been said to last 42 days (Smit, 1963; Hurxthal, 1979), and the limited information collected from Tsavo West National Park agrees with this figure. Hatching success could only be measured in three nests, in which 80, 86 and 100% of the central eggs hatched successfully. Not surprisingly, none of the eggs in the outer rings ever hatched.

4.6 Young

Ostrich chicks were rarely seen and difficult to observe in Tsavo West National Park for several reasons. First, the high rate of predation on nests meant that only a small proportion of nests survived until hatching. Second, the study periods ended before the eggs in a number of nests were hatched. Third, ostriches with chicks tended to be timid; not only did they and their chicks avoid our vehicle when still a considerable distance away, but the tall vegetation made it difficult to see the chicks well enough to count them. Chick survival was probably low (Section 5.5).

Hurxthal (1979) has described in Nairobi National Park, the creching of chicks, as a result of which the chicks from many nests in an area were to be found in one large flock being cared for by only two adults. The same

Plate 4.5   A pair of ostriches leads away a recently hatched clutch of chicks through concealing vegetation.

process probably occurred in Tsavo West National Park but no data are available.

Those chicks which survived grew rapidly, and had reached full height – if not their adult weight – within a year. They then made up the juveniles referred to in Section 3.1. At this age, they were no longer accompanied by adults, but moved around as a compact peer group, being much closer together than adults usually were. Their plumage was grey-brown and still clearly juvenile. The first signs of darker body feathers and paler wing feathers were detectable among the males.

By the age of 2 years, the young were easily sexed; however, the females were not easily distinguished from older females in the field. The males, on the other hand, were clearly distinguishable from all other age and sex categories, with their adult male plumage only partially complete. These immatures tended to move around in groups that were smaller and less compact than those of the juveniles.

By 3 years of age, the females could no longer be distinguished from older females. A small proportion of males was still obviously non-breeding, with their very pale necks, relatively drab black, rather than jet black, plumage and unconfident manner. It was presumed that they were 3-year-olds. No older categories were recognizable at the distances we observed the birds.

There are no data regarding the longevity of wild ostriches. Those in captivity are reported to live for at least 40 years (Smit, 1963). There are no indications that any of the birds were post-reproductive.

Underside of wing

5 # Ecological Aspects

Martial eagle predation

One of the aims of this study was to quantify the reproductive advantages and disadvantages of different breeding strategies. The benefits are often measured in later chapters in terms of numbers of eggs laid. Set against these, two of the most prevalent costs are the loss of eggs and the death of birds, and in this chapter I endeavour to quantify some of the effects of the natural hazards that ostriches face in Tsavo West National Park.

Predation on nests was a highly significant hazard. The great majority of ostrich nests did not produce any chicks, mainly because of predation. In order to understand the scale and nature of this predation, it is necessary to outline the different types of nest predator and the ways in which they wreak their effects. An attempt was made to quantify the scale of the threat they posed by a combination of observation and experimentation. The ostrich's evolutionary response to the combined effect on their nests of predators and heat was also investigated. So was the way in which adult birds protected themselves against such predators as lions. Finally, various seasonal factors were examined to determine to what extent they were responsible for the significant amount of variation in the detailed outcome of each breeding attempt.

5.1   Nest Predators

a) *Egyptian vultures.* Egyptian vultures (*Neophron percnopterus*) detected

Plate 5.1   An Egyptian vulture, rarely seen in Tsavo West National Park except at ostrich nests.

ostrich nests from the air during daylight hours. The shiny white ostrich eggs are conspicuous from above if they are not covered by an adult. The Egyptian vulture is one of the smallest and least abundant of the six local vulture species, but it is the only one capable of breaking open intact ostrich eggs in order to feed on the contents. Other vulture species, and other birds of prey, were seen to be unable to break the eggs open.

As described by Alexander (quoted by Andersson, 1856), van Lawick-Goodall (1968), Boswall (1977b) and Brooke (1979), Egyptian vultures have developed the technique of throwing stones at ostrich eggs to shatter the shell. This behaviour was widely used by Egyptian vultures in Tsavo, but they used a wide range of missiles, including quartz lumps, which appeared to be most effective; however, since these were not abundant in the study area lumps of wood and large disused snail shells were also used. Clearly they were not very satisfactory because usually several were found in a nest that had been attacked by Egyptian vultures. The snail shells, in particular, were liable to shatter before the ostrich shell, but were sometimes useful if full of soil. Being bright white, they were presumably easy to find compared with stones. They might be collected at some distance from the nest; on one occasion, a bird was seen flying to an ostrich

Plate 5.2   A selection of the missiles (rocks, snail shells and bits of wood) used by Egyptian vultures to try to break open ostrich eggs. A half ostrich eggshell shows the scale.

nest from a distance of more than 150 m carrying a snail shell which it then proceeded to throw at the eggs.

Each missile was thrown with a downward flick of the beak while the bird stood on the ground close to the egg; it was not dropped from the air (Boswall, 1977a; Brooke, 1979). It appeared that a number of direct hits were required – even with the best missiles – to break the 2-mm thick eggshell, and this process often took several minutes. The eggs were broken *in situ*, not moved. Those nests destroyed by Egyptian vultures could be identified by the mixture of shell remains and missiles inside the nest scrape.

An Egyptian vulture cannot prey on a nest if an adult ostrich is in attendance. It is noteworthy that none of the outer ring of eggs around incubating ostrich nests was ever destroyed by an Egyptian vulture (despite their conspicuousness from above), presumably because of the presence of one or other adult. Ostriches clearly recognize that Egyptian vultures pose a threat to their nests. On two occasions, an ostrich (once a male, once a female) was observed to run straight back to its nest (something ostriches never usually do) on seeing an Egyptian vulture land near it.

Plate 5.3 A small nest destroyed by an Egyptian vulture; note the bird's feather, several land snail shells which proved unsuccessful missiles, and the small quartz rock which eventually broke the egg.

These vultures fly by day only, and therefore are only a danger to unattended eggs during daylight hours. They did not appear to be common in Tsavo, and were rarely seen away from ostrich nests. No information exists regarding the local distribution and movements of Egyptian vultures; nor do we know how high they fly and at what times of day. All these factors are relevant in assessing the threat which they pose to unatttended ostrich nests.

b) Black-backed jackals. Black-backed jackals (*Canis mesomelas*) destroyed a number of ostrich nests at night, when most jackal activity takes place (and also when our time-lapse photography was of no use). The jackals presumably discovered the nests by smelling either the eggs or the birds, but it is not known at what range they are able to do this. Certainly, ostrich nests are virtually invisible to ground predators except at very close range (less than about 15 m).

Although the method by which the jackals broke the eggs is unknown, it is likely that they did so by rolling the eggs against each other – nests destroyed by jackals usually contained one or more unbroken eggs, sometimes displaying percussion fractures. Most of the remains of broken eggs

Plate 5.4   A young family of Black-backed jackals examine an ostrich egg but appear uncertain what to do.

were scattered just outside the nest edge, but possibly were pulled there after the egg had first been broken in the middle of the nest. Jackals are far too small to be able to carry intact eggs away from the nest site; but eggs were sometimes found scattered up to 30 m or more from the nest.

There is no information on the extent to which ostriches are able actively to defend their nests against jackals. During the daylight hours, they almost certainly could, but it is possible that at night a pair of jackals could force a sitting ostrich to stand, so enabling them to get at the eggs. We did record by time-lapse photography instances of birds sitting on a nest at dusk that next morning had been destroyed. Although jackals had fed at three of these nests, so too had hyaenas, and it was impossible to determine which was responsible for destroying the nest. Both species are scavengers and would be attracted to a nest by the smell of broken eggs.

We recorded little direct information on the abundance or distribution of jackals, other than records of animals encountered by chance. These show a marked increase in the numbers of jackals sighted in 1979 compared with the two previous years. However, because most sightings of jackals were on tracks in the late afternoon or early evening, this may be a reflection only of this largely nocturnal animal's use of tracks during daylight hours, rather than of population density. Certainly, the propensity to emerge in daylight changed markedly even within single seasons. On the other hand, there was a rodent (*Arvicanthis*) infestation in 1978, and therefore an increase in jackal numbers by 1979 might be expected.

c) *Hyaenas.* The spotted hyaena (*Crocuta crocuta*) and the striped hyaena (*Hyaena vulgaris*) were rarely heard or observed in the study area. Hyaenas of unknown species were involved in a proportion of nest losses, as evidenced by their footprints. Nest destruction by hyaenas occurred at night and like the jackals, hyaenas probably found the nests by smell. Hyaenas probably have little difficulty in breaking ostrich eggs with their teeth, and may even be able to carry eggs away intact. Eggs were often removed completely from the vicinity of a nest.

Although adult ostriches are probably able to keep a hyaena away from their nest during daylight hours, they are unlikely to be able to defend the nest against even a single hyaena at night. We recorded no information on the abundance or distribution of hyaenas.

d) *Lions.* Lions (*Panthera leo*) almost certainly destroy ostrich nests if they are encountered, although we recorded no cases of their doing so. However, an egg given to a lioness in Tsavo was so efficiently demolished that we felt certain that she had done so before. She held the egg down with her forepaws, made a hole at the top with her canine teeth, and lapped up the contents through the hole with her tongue, spilling almost none.

Ostriches perform a distraction display which became more pronounced as incubation progressed. It is similar to the broken-wing display of many small ground-nesting birds, involving the ostrich scuttling away with wings low and flapping irregularly, and then collapsing on the ground. It is not known how effective this is at luring large predators such as lions away from a nest. Lions seek their prey mainly by sight (Schaller, 1972); the concealment of ostriches' nests and the birds' general wariness are ostriches' main defence against predation by lions.

5.2   The Causes of Nest Destruction

Data were collected on the causes of the destruction of ostrich nests; Table 5.1 summarizes these data for all 57 nests. The land-based carnivores – hyaenas and jackals – were responsible for destroying 40% of the nests. In addition, they were also involved in the destruction of eight other nests, after fire had removed the cover concealing five nests and after Egyptian vultures had plundered three others. In two nests, the apparently

Plate 5.5   A lioness tackles an ostrich egg with her paws, teeth and tongue.

accidental breakage of one and 16 eggs could not be explained. Three nests were abandoned, two after major disturbances.

Table 5.2 gives information on the fate of each nest found. Nest destruction took place at any time, ranging from a day after laying to the day of

Table 5.1

The causes of ostrich nest destruction

	No.	%
Carnivores		
jackals	6	11
hyaenas	8	14
hyaenas and/or jackals	9	16
Egyptian vultures	6	11
Fire	5	9
Accidents	2	4
Abandoned	3	5
Hatched successfully	5	9
Still surviving at the end of the study periods	13	23
Total	57	

Table 5.2

The fate of ostrich nests

Year[1]	Nest no.[2]	Approximate age when found (days)[3]	Approximate age when destroyed (days)[4]	Reached incubation?[5]	Hatched?[6]	Cause of loss[7]
1973	A	8	36	Yes	No	Fire, then jackals/hyaenas
1977	1	16	26	No	—	Abandoned, then Egyptian vultures
	2	4	—	Yes	Yes	—
	3	1	16	No	—	Jackals
	4	3	6	No	—	Hyaenas
	5	8	19	No	—	Jackals/hyaenas
	6	19	21	No	—	Jackals/hyaenas
	7	8	19	No	—	Jackals/hyaenas
	8	1	9	No	—	Egyptian vultures
	9	4	10	No	—	Jackals/hyaenas
	10	7	8	No	—	Egyptian vultures
	11	11	16	No	—	Hyaenas
	12	Destroyed	8	No	—	Probably jackals
	13	9 weeks	—	Yes	Yes	—
	14	Incubating	—	Yes	?	Surviving
	15	1	—	?	?	Surviving
	16	Incubating	—	Yes	?	Surviving
	17	Incubating	—	Yes	?	Surviving
	18	1	12	No	—	Egyptian vultures
	19	Destroyed	7	No	—	Jackals/hyaenas
1978	1	6	46	Yes	No	Fire, then jackals/hyaenas
	2	3	14	No	—	Fire, then jackals/hyaenas
	3	10	18	No	—	Vultures, then jackals/hyaenas
	4	3	53	Yes	No	Jackals/hyaenas
	5	10	—	Yes	Yes	—
	6	Incubating	Incubating	Yes	No	Fire, then jackals/hyaenas
	7	14	21	No	—	Jackals
	8	7	12	No	—	Jackals/hyaenas
	9	Incubating	Incubating	Yes	No	Jackals/hyaenas
	10	12	17	No	—	Jackals
	11	Incubating	—	Yes	?	Surviving
	12	After hatched	—	Yes	Yes	—
	13	9 weeks	9 weeks	Yes	Yes	Hyaenas as hatching

Table 5.2

Continued

Year[1]	Nest no.[2]	Approximate age when found (days)[3]	Approximate age when destroyed (days)[4]	Reached incubation?[5]	Hatched?[6]	Cause of loss[7]
	14	9 weeks	—	Yes	Yes	—
	15	14	—	Yes	?	Surviving
	16	Destroyed	21	No	—	Jackals
	17	Destroyed	16	No	—	Hyaenas
	18	15	17	No	—	Vultures, then jackals
1979	1	17	22	No	—	Jackals/hyaenas
	2	10	28	Yes	No	Hyaenas
	3	3	19	No	—	Vultures, then jackals
	4	1	—	Yes	?	Surviving
	5	7	13	No	—	Egg broken, then abandoned
	6	12	44	Yes	No	Fire, then jackals
	7	1	13	No	—	Hyaenas
	8	4	—	Yes	?	Surviving
	9	7	11	No	—	Jackals
	10	15	27	No	—	Unknown accident
	11	Incubating	—	Yes	?	Surviving
	12	9	11	No	—	Hyaenas
	13	16	—	Yes	?	Surviving
	14	9	19	No	—	Abandoned
	15	1	6	No	—	Abandoned
	16	9	—	?	?	Surviving
	17	4	—	?	?	Surviving
	18	2	—	?	?	Surviving
	19	Destroyed	2	No	—	Hyaenas

Notes: The ages of nests when found (column 3) were estimated either from the number of major hens' eggs in the nest or from the graph shown in Fig. 4.7. Day 1 was the day on which the first egg was laid.

Nests where it was not known whether or not incubation was reached (? in column 5) were those discovered late in the season and still at the laying stage upon our departure. Similarly, ? in column 6 indicates that the nest was still extant upon our departure. The dash, —, in column 6 indicates that the nest obviously did not hatch because it failed to reach the incubation stage.

The dash —, in column 7 indicates that the nest hatched successfully.

hatching. Egyptian vultures, if they destroyed a nest, did so earlier (mean 13.8 days) than terrestrial carnivores (mean 19.7 days). This difference was caused mainly by the fact that Egyptian vultures did not destroy any incubating nests (probably because either the male or female ostrich was almost always in attendance; see Fig. 4.8), whereas jackals and hyaenas did; the mean age of pre-incubation nests destroyed by jackals and hyaenas was 14.1 days.

Predation on nests was usually, but not necessarily, an all-or-nothing affair. Three nests were known to have lost an egg to Egyptian vultures at an early stage, yet survived for at least another week. However, the vultures tended to return at a later date to plunder further a nest which they had discovered, and the smell of the remains of any broken eggs attracted mammalian predators. Jackals did not usually destroy an entire nest in one night, and the adult ostriches sometimes returned and endeavoured to continue with it; however, the jackals almost always destroyed further eggs on subsequent nights. Only in one case did a nest survive this repeated predation over a few days; it lost eight eggs in total, and resulted in the smallest incubating clutch in the study (14 eggs: see Table 4.3).

5.3   Nest Predation Risk

The ostrich nesting system appears adapted to a high risk of predation, and it is important to be able to quantify this predation risk, first during the laying period and later during the incubation period. Quantification is not straightforward, because each of the possible methods has different inherent biasses.

The risk run by *unattended* ostrich nests could to some extent be quantified by experiments. Dummy nests were set up, and the time elapsing until their destruction was recorded. Altogether, 22 such artificial nests were established, containing a mean of 9 eggs (range 3–23 eggs). Because an abundant supply of real eggs was unavailable, most nests consisted of a mixture of a few real intact eggs, some empty shells, and some realistic fibreglass dummy eggs or half-eggs – some empty eggs containing a hen's egg to compensate (partially) the predator for its efforts. These nests were set up at infrequent intervals (2 days–3 weeks) during all three study periods, scattered over an area of about 200 km². The sites were chosen to be representative of where normal ostrich nests were found. That they were indeed reasonably representative was suggested by the facts that an ostrich started laying regularly in one of the dummy nests, and that another nest was found and damaged by a group

Table 5.3

The survival and destruction of dummy nests

Dummy nest no.	Year	No. of eggs in nest	Survival period (days)	Cause of destruction or termination
D.1	1977	23	6	Egyptian vultures
D.2	1977	6	3	Egyptian vultures
D.3	1977	6	4	Egyptian vultures
D.4	1977	5	14	Egyptian vultures
D.5	1977	4	1	Jackals
D.6	1977	3	>30	Survived to criterion
D.7	1977	4 → 10	>30	but ostrich had started laying in it
D.8	1977	6	(>13)	Terminated by observer
D.9	1977	5	11	Egyptian vultures
D.10	1978	5	5	Egyptian vultures
D.11	1978	5	8	Fire
D.12	1978	6	18	Egyptian vultures
D.13	1978	5	>30	Survived to criterion
D.14	1978	5	29	Egyptian vultures
D.15	1978	15	1	Egyptian vultures
D.16	1978	15	1	Egyptian vultures
D.17	1978	15	6	Probably jackals
D.18	1978	15	(>9)	Terminated by observer
D.19	1979	13	4	Destroyed by ostriches
D.20	1979	5	>30	Survived to criterion
D.21	1979	13	>30	Survived to criterion
D.22	1979	14	28	Jackals
		$\bar{x} = 8.8$		

Notes: Number destroyed within 7 days = 9 (41%); % of destructions caused by Egyptian vultures = 10/15 (67%).

of ostriches which moved the eggs and broke one of the empty shells. The dummy nests were visited at intervals of 2 days or so and any destruction recorded.

A large proportion (41%) of these dummy nests was destroyed within the first 7 days (Table 5.3), and Egyptian vultures were mainly responsible (67%). Fire was responsible for the destruction of one dummy nest, jackals for three and adult ostriches for another. Given the high probability of early destruction, a surprisingly high proportion of dummy nests survived for more than 30 days, after which they were removed. In other words, a dummy nest tended either to be destroyed quickly or else not to be destroyed at all. The reasons for this are unknown – it could result from particular patterns in vulture search methods, from a patchy distribution

Plate 5.6 The remains of a large dummy nest after destruction. Note the broken shells and fibreglass eggs and half-shells.

of vultures capable of destroying eggs, or from other similar causes. The eggs clearly remained a source of food for vultures for several weeks.

Only four dummy nests were set up in 1979 mainly because they were not being destroyed. This meant a shortage of eggs, and it would have been undesirable to have too many artificial nests in the environment at any one time. The predation rate by Egyptian vultures both on real and on dummy nests in 1979 was lower than in the previous two years. No information is available on vulture numbers, because they were so rarely seen.

Using the dummy nests, we tried to determine whether large unattended nests are more vulnerable than small ones. Table 5.4 shows that the rate of destruction of large dummy nests was almost double that of small dummy nests. The difference, which was not significant, was presumably due to the greater number of eggs in the larger nests being more conspicuous.

Real ostrich nests were not left unattended for very long periods except at the early stage of laying. The presence of a bird at a nest reduces the visual conspicuousness of that nest, particularly from above. However, it may also increase the nest's likelihood of discovery by terrestrial predators, which hunt by smell or by observing the behaviour of adult ostriches. Overall, 42% (22 of 53) of nests whose fate was known survived

Table 5.4

The survival of large and small dummy nests

	Large nests (13–23 eggs)	Small nests (3–6 eggs)
Number of nests	8	14
Median survival period	6 days	13.5 days
Probability of destruction within 7 days	63%	29%
Mean number of days of survival per destruction	14	25

Note: Data from Table 5.3.

until incubation, but it must be stressed that this figure was biassed by the methods used to find nests. In most cases, we were led to the nest by an adult ostrich, and were therefore more likely to discover nests that survived rather than ones that were destroyed. Similarly, a variety of other biasses arise with the results shown in Table 5.5 of the various methods for calculating the daily nest destruction risk during the laying period.

Method A was to take all the nests of which we had knowledge (including all the incubating nests), and to divide the total number of nest break-ups during the laying periods by the total number of days in those periods – we assumed that for all incubating nests the laying period had

Table 5.5

The destruction risk for nests during the laying period and during the incubation period

Method	No. of nests for which data could be used	Destruction rate	Daily risk
Laying period			
A. All nests known	57	31 destructions in 952 laying days	0.033
B. All nests found or destroyed during laying period	46	31 destructions in 699 laying days	0.044
C. Nests found and observed during the laying period	39	27 destructions in 379 laying days	0.071
D. Dummy nests	22	15 destructions in 311 days at risk	0.048
Incubation period			
E. All incubating nests known	20	7 destructions in 469 incubating days	0.015

lasted 23 days, the mean figure given in Section 4.3. This gives a figure of 0.033, which is probably low because it is biassed by those nests which survived until incubation and until they could be discovered by us. Method B was to consider only all of those nests which were discovered (or found destroyed) during their laying periods and to calculate in the same way as above. This gives a figure of 0.044, which for the same reason as above is biassed in the same direction, but to a lesser extent. Method C was to take only those nests which had been discovered and were still surviving at the laying stage, and to calculate their destruction rate in the subsequent days following their discovery. This figure (0.071) might be biassed upwards if our visits caused any increase in the destruction risk of nests, e.g. if predators followed vehicle tracks. Method D was to use the data from the dummy nests, resulting in a figure of 0.048; this is only partially representative of normal nests because, unlike them, dummy nests were not attended; they therefore ran a greater risk from vultures but probably a reduced risk from mammals. The four figures calculated and shown in Table 5.5 do not differ to a very large extent, and we may therefore conclude that the risk that an ostrich nest will be destroyed on any day during the laying period is of the order of 0.06. This would correspond with a survival rate until incubation of 24% of the nests started. We did not collect any data on how the destruction risk altered during the 23-day laying period, because of the biasses in the chances of finding surviving *vs* destroyed nests.

The ostrich nests were not safe during the 6-week incubation period; indeed, 7 of the 12 nests (58%) in which incubation had begun and whose fate was known were destroyed before they hatched. For some of these nests, there is no accurate information as to which stage of incubation they were at, and therefore it cannot be determined whether their risk of predation varied during the course of the incubation period.

Table 5.5 shows the average predation risk calculated for all incubating nests in the same way as for during the laying period, i.e. the total number of break-ups divided by the total number of days. It gives a daily risk of destruction of 0.015 during incubation, which corresponds to the destruction of 46% of incubating nests durng the 42-day incubation period. There are probably three reasons why the daily risk of destruction during incubation was markedly lower than that during laying. First, attendance by adult ostriches was almost 100% (Fig. 4.8). Second, incubating birds were probably more willing to defend their nest against predators such as jackals against whom defence might be effective. Certainly, they left the nest later and moved less far away from it at the approach of our vehicle, and were more likely to exhibit a distraction display. Third, any nests in relatively vulnerable sites were more likely to have been discovered and

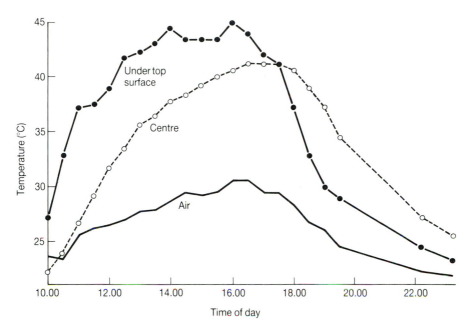

Figure 5.1   The temperatures at the centre and under the top surface of an ostrich egg exposed all day to the sun; the ambient air temperature is also shown.

destroyed already. Depending to some extent on how predators search, an older nest may on average be an intrinsically safer nest, solely because the less safe ones have already been eliminated.

5.4   Temperature

Predators are not the only hazard faced by ostrich eggs – high temperatures are another. Figure 5.1 shows the variation in temperature at a small unattended ostrich nest in full sun in Tsavo West National Park. The egg temperatures were recorded at 30-min intervals by means of thermometers which were stuck vertically down into a fresh ostrich egg; the ambient shade air temperature was measured some 30 m away. It can be seen that the central temperature of the egg rose more rapidly than did the shade air temperature, and reached a maximum of 41.1°C. This and three repeats of this experiment on different days gave a mean maximum central temperature of 40.8°C, which approaches the lethal temperature for domestic hen eggs (42.2°C: Lundy, 1969). However, the blastoderm itself invariably

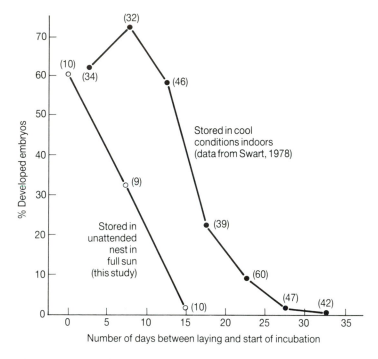

Figure 5.2 The viability of ostrich eggs after different storage periods and conditions before incubation.

floated buoyantly just under the upper surface of the shell, as was clearly seen whenever an ostrich egg was opened; therefore, since a marked temperature gradient could be felt across such large eggs, the blastoderm was in fact exposed to appreciably higher temperatures. Also shown in Fig. 5.1 are the temperatures recorded just inside the upper surface of the shell of the same egg, and it can be seen that they rose as high as 45°C, which is dangerously high. Inadequate data exist on the heat tolerance of wild ostrich eggs. However, Fig. 5.2 shows the viability of domestic ostrich eggs after being exposed to full sun in an unattended nest for periods of 0, 7 and 15 days (Bertram and Burger, 1981a). Their viability declined sharply with increasing length of exposure, and that of eggs exposed to the sun was considerably lower than that of eggs stored in cool conditions (Swart, 1978).

Thus it is important for adult ostriches to attend their nests not only to deter predators, but also to prevent the eggs from overheating. Attendance by adult ostriches during the early stages of laying was mainly during the hottest middle part of the day (Fig. 4.3), when the eggs were most likely

Plate 5.7 Eggs used in experiments to test the vulnerability to predators of brown (left), normal white (centre) and red (right) eggs.

to be affected by heat stress. It was not possible to quantify the threat of overheating, but it must have been appreciable.

A series of experiments was carried out to demonstrate the interaction between these two threats (Bertram and Burger, 1981b). It seemed possible that the high predation rate by Egyptian vultures might have been due in part to the conspicuousness of the shiny white eggs. In order to test this, 12 ostrich eggs were dyed a medium brown colour and their vulnerability to predation was compared with that of 12 normal white eggs. Because there was a shortage of whole eggs, six of the latter were ostrich eggshells containing a hen's egg to reward any predator. The white and dyed eggs were placed alternately on the bare ground more than 500 m apart. They were visited usually every other day, so that the time elapsing until their destruction could be recorded. In addition, a further two eggs were dyed bright orange and also put out; these were designed to be as conspicuous as the white eggs but as unfamiliar to predators as the brown ones. The white eggs were certainly more conspicuous to the human eye.

Three eggs (one of each colour) were removed by mammals and so are omitted from further consideration. Of the camouflaged brown eggs, six of 11 (55%) survived for more than 10 days, whereas only two (17%) of

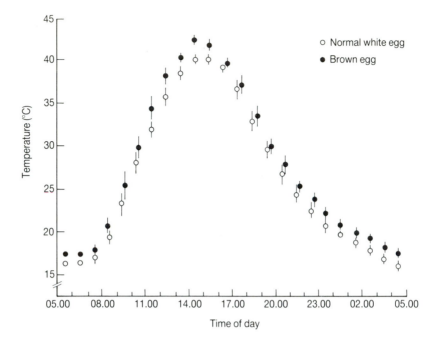

Figure 5.3 The central temperatures reached by normal white and artificial brown ostrich eggs during 24 h exposure to sunlight.

the 12 conspicuous eggs did. The differences in survival times were statistically significant ($P < 0.05$, Mann-Whitney test, $n_1 = 11$, $n_2 = 12$). The camouflaged eggs survived roughly three times as long as the normal eggs (3.5 as opposed to 10 days). The orange eggs were destroyed as quickly as the white ones, which suggests that, despite the small sample, the eggs dyed brown were in fact concealed and not just made unfamiliar (Bertram and Burger, 1981b).

It was possible to demonstrate that white eggs, while suffering a greater predation risk, had a corresponding gain due to their whiteness. The temperatures reached by eggs in an unattended artificial nest at Oudtshoorn in South Africa were measured. The nest contained 12 eggs, of which 10 were white and two were coloured medium brown with brown, black and green crayons. Thermistors were inserted from below into the centres of one white and one similar-sized (1365 g) brown egg, and the temperatures reached were recorded at 10-min intervals over a 24-h period (see Fig. 5.3). Two other less complete experiments in Tsavo on different days yielded comparable results. The average maximum central temperature reached was 3.5°C higher for the brown eggs (43.3°C) than for the

Plate 5.8   A Martial eagle, an aerial predator on ostrich chicks.

white eggs (39.8°C). Measurements taken with an infra red field thermo-
meter clearly showed that the high temperatures of the coloured eggs were
not an artefact of the colouring process but were due to the colour *per se*
(Bertram and Burger, 1981b).

Thus in evolutionary terms, ostrich eggs have traded an increased
predation risk for greater safety from overheating; white eggs may be more
vulnerable but they do not get so hot.

5.5   Predation on Birds

We collected little information on chick mortality. It is almost certainly
very high, since the chicks are obviously vulnerable and predators are
abundant. Hurxthal (1979) reported that in Nairobi National Park, chick
survival to the age of 1 year was of the order of 10–15%. It is possible to
calculate that the survival rate of chicks in Tsavo must have been of
roughly the same order, given the number of successful nests. It is not
known whether the large variation in juvenile numbers between years
(Table 3.1) was caused by differences in nesting success or by differences
in chick survival rate, though the former is more likely.

At first, when chicks hatch, they move as a group with the two adults who hatched them. Despite predation, the mean group size increases due to the merging of several of these groups (Hurxthal, 1979). It is likely that being in a group improves an individual chick's chance of survival by reducing the likelihood of its being a victim of a predator who attacks the brood. This protection by 'dilution' (Bertram, 1978) operates provided (1) that not all of the prey within a group are taken by the predator and (2) that the rate or success of predator attack does not increase at a greater rate than group size. The first point is certainly true: there are numerous reports of predators such as raptors and hyaenas killing one or a few of the ostrich chicks in a group (e.g. J. D. Bygott and T. Corfield, pers. comms), and the gradual decline observed in the size of chick groups (Hurxthal, 1979) denotes partial escape from whatever predator is at work. The second point is more difficult to demonstrate, e.g. there are no available data regarding the conspicuousness or vulnerability of ostrich chick groups in relation to the size of the groups. In Tsavo, when the chicks are small, they can only usually be detected among the vegetation by the behaviour of the adults accompanying them. It is unlikely that a group of 20 chicks would attract twice as many predation attempts as a group of 10 chicks. However, data are needed on this point, and also on the success rate of predators attacking chick groups. It may well be that adult ostriches are less good at defending 20 than 10 chicks; it needs to be determined how good they are. But unless a group of 20 chicks suffers two or more times as many successful predation attempts as a group of 10 chicks, the individuals in the larger group benefit from being in such a group.

The same protection against predation by being in a group operated not only among small chicks but also among older ostriches. Juveniles travelled in compact groups, and adults were often with companions. An investigation was undertaken into whether or not adult ostriches benefit from the presence of their companions when they feed as a group instead of on their own (Bertram, 1980). A bird pecking at food has its head down among the vegetation and cannot see what is going on around it. In Tsavo, adult ostriches raised their heads at intervals and stood motionless, apparently looking about them. After a few seconds, their head was lowered and feeding resumed. In 1977, I recorded how often and for how long males and females raised their heads when they were feeding near to different numbers of companions. The details of the methods, results and conclusions are given by Bertram (1980); the following is a summary.

There was a decline in a bird's 'percentage vigilance' (the proportion of the time its head was up) as the number of birds nearby increased (Fig. 5.4). For birds feeding alone, the median percentage vigilance was 34.9, for

Plate 5.9   A male ostrich alternates bouts of feeding with its head down among the vegetation, with bouts of vigilance with head held high.

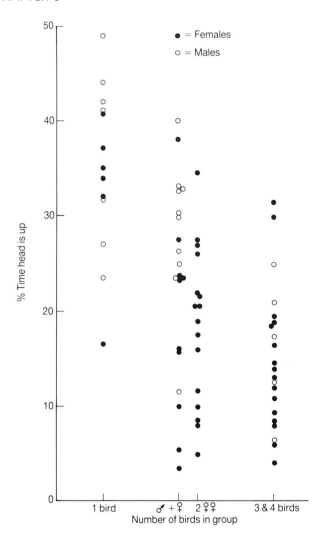

Figure 5.4   The percentage vigilance of ostriches when feeding in groups of different sizes.

birds in groups of two it was 22.9, and for birds feeding in groups of three or four it was 14.0. The differences were due mainly to the fact that the birds in larger groups raised their heads *more often*, rather than because they kept them raised for longer. It did not appear to be a result of the birds gathering at richer feeding places.

Male ostriches were more vigilant than females (Fig. 5.4). The difference was caused mainly by their keeping their heads up longer each time they were raised, rather than by their raising them more often. A female feeding with a single companion was equally vigilant whether that companion was male (and therefore himself more vigilant) or female (Fig. 5.4); thus the presence rather than the vigilance of the companion was what influenced her own vigilance.

When two or more birds were feeding together, they raised and lowered their heads independently of one another; there was no indication that they organized their vigilant periods in any way, such as taking turns at raising their heads, or one having its head down while the other had its head raised. The amount of time that elapsed between raises of the head varied, and therefore it was impossible (for either myself or a predator) to predict when the head would next be raised.

Lions were probably by far the most important predator of adult ostriches in Tsavo; the other predators were too small or too scarce to have any significant impact. One instance of a lion feeding on an adult ostrich was reported in the Tsavo study area (J. Cheffings, pers. comm.) and a few other records are available (Schaller, 1972; Rudnai, 1974; Pienaar, 1969). I have observed lions unsuccessfully stalking ostriches, both in Tsavo and in the Serengeti National Park. By making certain assumptions about lion predation on adult ostriches, we can to some extent quantify the benefit which ostriches gain by feeding in groups (Bertram, 1980): a lion must stalk to within a short distance of an ostrich with its head down in order to have a chance of catching it; a bird with its head up is invulnerable; all birds in a group are invulnerable when one member has its head up; when a lion attacks, its victim (if any) is a random member of the group; and a small group of ostriches does not attract more predation attempts than a single bird.

Figure 5.5 shows the changes in relative vulnerability with group size. Line 1 (taken from Fig. 5.4) shows the mean proportion of time that any particular individual bird has its head down. Line 2 shows the group's vulnerability: it is the proportion of time that all the heads in the group are down and there is no member vigilant. It can be seen that the group's vulnerability changes little with group size; although there are more eyes to look with, each pair of eyes is used less. The predator's chance of success is thus little influenced by group size. Line 3 shows each bird's individual vulnerability (this is line 2 divided by the number of birds in the group). It is clear that a bird's individual vulnerability decreases rapidly as group size increases, and that its first companion is much the most important. Protection is gained almost entirely by diluting the predator's success (Bertram, 1978), not by reducing it.

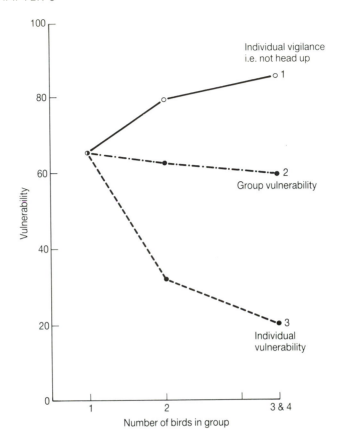

Figure 5.5   The relative vulnerability of ostriches in relation to group size.

5.6   Seasonal Factors

It is possible to gain insight into some of the ecological factors influencing ostrich nesting by examining both the differences between different years and the variety among individuals.

a) Season of nesting. An important aspect of ostrich breeding is the season of nesting. Figure 5.6 shows the dates at which ostrich nests were started in each of the 3 years of the study. It should be noted that almost all of the nests of which I had any knowledge are included, and that therefore a few of the starting dates were not more precisely known than to within about 2 weeks.

Because no observer was present in the study area before July each year,

Plate 5.10   A group of juveniles. By grouping, the individual's risk of being preyed upon is considerably reduced, through dilution and the group's vigilance.

Fig. 5.6 does not record nests begun before that time; however, had nests been started before July, it would have been possible to detect them during their incubation stage. No chicks were seen which could have come from nests started before July. Some nests were still being started at the time of our departure.

Figure 5.6 shows that there was a reasonably well-defined breeding season in the sense that all nests were started within 4 months of one another, during the months of July–October. The relation between these months and the average monthly rainfall is shown in Fig. 5.6. It is clear that in this part of Tsavo, ostriches only nested during the dry season.

Comparisons between the 3 years show that the immediate climatic conditions appeared to have remarkably little effect on the timing of ostrich breeding. The median starting dates were 31, 16 and 24 August for 1977, 1978 and 1979 respectively. Although there were considerable differences between years, regarding both total amount of rainfall and its distribution pattern, none the less the temporal distribution of ostrich nests in each year was remarkably constant.

b) *Food supply*. It was impossible to measure the food supply directly, but there were clear differences between the vegetation in the three dry season

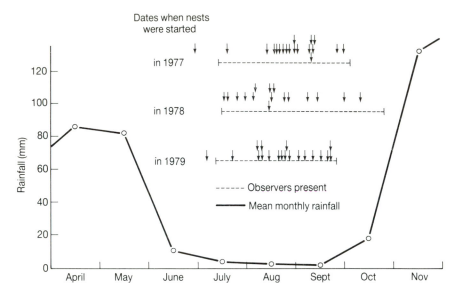

Figure 5.6   The dates at which ostrich nests were started in each of the study seasons, and average monthly rainfall.

study periods. In 1977, following low rainfall in previous years, the ground vegetation was sparse, very dry and fairly low, and none of the bushes had leaves. Food was probably in short supply. In 1978, there had been high rainfall early in the year; as a result, the ground vegetation was thick and tall, and most bushes bore leaves. The great majority of the area and of the vegetation was burnt off by a vast fire in early August, and so during most of the study period cover was scanty. Food appeared to be moderately abundant due to the rapid re-growth of much of the ground vegetation. In 1979, there had again been high rainfall during the prolonged rainy season. The ground vegetation was very thick and tall and still fairly green, and all bushes were still in leaf. The dry season was not as hot or arid as in previous years, and as a result the vegetation remained reasonably green throughout the dry season. Food appeared to be abundant for the whole period.

c) *Feeding rate.* Only in 1979 was it possible to obtain data on the feeding rate, by measuring the proportion of the time that birds spent feeding (either feeding or nibbling: see Section 2.2) at different stages during the 3-month study period. Figure 5.7 shows that there was no significant trend in the proportion of time spent feeding during this period. In any case, it is difficult to interpret data on time spent feeding: more time spent feeding

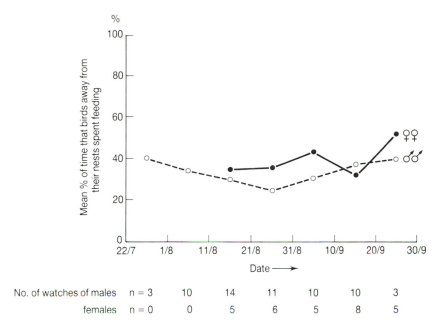

Figure 5.7   The proportion of time spent feeding by males and females when not at the nest during the 1979 study period.

may indicate a higher food intake, or it may indicate that food is more difficult to obtain and thus that less of it is obtained per unit of energy expended. Under the prevailing observation conditions with distant birds and abundant ground cover, it was not possible to quantify their food intake in terms of their pecking rate or of the size or frequency of passage of a food bolus down the neck. Subjectively, there were no very obvious differences between years in the proportion of time spent feeding, but we have no quantitative information to support this impression. There were obvious differences in food abundance, but again there is no information on the quality of this food.

d) *Numbers of eggs.* We tried to determine whether differences in food abundance between years resulted in differences in the numbers of eggs laid in those years: the results were inconclusive. The total number of nests found in each year was remarkably constant at 18 or 19 (Table 5.6). Obviously, this reflects in part the amount of effort expended searching, which was roughly similar from one year to the next. The total numbers of eggs found each year showed more variation (Table 5.6). In each year, eggs were still being laid at the time of our departure; the total numbers of eggs found

Table 5.6

The numbers of nests and eggs in 1977–79

	1977	1978	1979
Overall			
Nests found	19	18	19
Eggs found	308	397	247
In 7 territories			
Nests found	10	8	10
Eggs found	107	165	135

reflect partly the different durations of observations. In 1979, we concentrated on the nests of certain known males; this excluded searching for and finding large incubating nests further afield, as had been done in the two previous years. Therefore, the numbers of nests and eggs laid in the territories of the seven adult males for which data were available for each year are shown in Table 5.6. None of these nests was still being laid in upon our departure. Table 5.6 shows that in 1978, 54% more eggs were found in these males' territories than in 1977. However, there is so much variability between events in different males' territories that this increase (based on only seven territories) may not be representative of the local population.

e) Egg size. There were indications that the average size of the eggs laid in different years altered. Figure 5.8 plots the frequency distribution of the weights of all the ostrich eggs which were weighed during the laying period at each nest for each season. The eggs that were weighed only during the incubation period have been excluded, because there was a much lower proportion of these in the third season, and because ostrich eggs lose weight at a mean rate of between 3.0 and 4.3 g per day during the incubation period (Bertram and Burger, 1981a). Eggs also lose weight during the laying period, but at a slower rate (1.7–2.9 g per day: Bertram and Burger, 1981a), and there was no significant difference between years in the age of eggs when weighed during the laying phase.

It can be seen from Fig. 5.8 that the median fresh egg weight increased each year, and that in 1979 it was 102 g (6.5%) heavier than in 1977. The samples were reasonably large. This increase might be a reflection of the more abundant food available in the later seasons, but further information suggests that it may not be. If it were, we should expect the weights of the eggs of known individual birds to be greater in 1979. However, Table 5.7 indicates that for the three females for which there are sufficient data, their average egg weight did not increase in 1979, and indeed in one case decreased significantly. Part of the increase in the average egg weight in

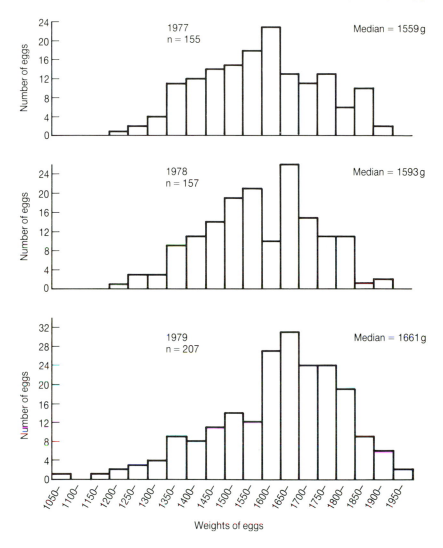

Figure 5.8   The frequency distribution of egg weights in each study season.

1979 was due to the large number of eggs laid by one particular female (Lame), who laid a large number of unusually large eggs in 1979; she was certainly laying in the region in 1978, but there is no information as to how many eggs she laid or of what size. The factors determining egg size and its relationship to food abundance are thus unclear. The great variation in egg size even within a season is most striking.

Table 5.7

The mean egg weights of three individual female ostriches, 1977–79

	Mean egg weight (and no. weighed)		
Bird	1977	1978	1979
Bwing	1399 (8) (±53)	1373 (11) (±52)	1351 (14) (±82)
Goldie	1509 (6) (±63)	? (—)	1512 (5) (±22)
Jo	? (—)	1758 (13) (±61)	1678 (16) (±43)

f) *Predation rate.* As indicated above, the predation rate on nests was high. By taking all those nests which were found during the laying phase and whose fate was known, and measuring their survival period, we were able to compare the predation rates between years. Since the observers left the study area before many nests had hatched, and this date was later in 1978 than in 1977 and 1979, we have used 42 days as the period of survival for nests reaching incubation; this figure is made up of 23 days of laying, plus 46% of the incubation period of 42 days (since about 46% of nests which started incubation survived until hatching). Table 5.8 shows the mean survival period of nests in each of the 3 years. Despite the great variation within years, it is clear that the nests observed survived less well in 1977 than in the two subsequent years. The reasons for the lower predation rates in the later years are not clear, and may be due largely to chance. They are not related in any obvious way to climatic differences, which overall appeared to have relatively little effect on ostrich nesting. They may be related to the availability of cover; but even within the 1978 season, the fire which removed almost all the vegetation from almost the whole of the study area had no detectable effect on ostrich nesting.

Table 5.8

Nest survival, 1977–79 (for nests found during the laying period)

	1977	1978	1979
Proportion of nests which survived until incubation	1/14	4/12	5/15
Mean survival period (days)	14.7[a]	24.7[a]	22.9

[a] $P < 0.05$.

Table 5.9

The survival periods of early- and late-started nests

	1977	1978	1979	1977–79
Median starting date of all nests	31 August	16 August	24 August	
Mean survival period of nests started before the median date that year	17.9	26.8	29.5	24.8
Number of nests included	7	5	8	20
Mean survival period of nests started after the median date that year	9.8	25.2	26.0	19.7
Number of nests included	4	5	2	11
Significance of the difference in survival periods	N.S.	N.S.	N.S.	N.S.

Notes: Only nests discovered during their laying period are included; those discovered during incubation would have biassed the sample because they were the more successful ones.

Nests begun before our arrival are not included, because they too would have biassed the sample by being the more successful ones.

Nests begun within 23 days of our departure are not included because they would have biassed the sample. Although we would have known of the destruction of these nests, we would not have known if they had survived until incubation.

None the less all these nests were included for calculating the median starting date for that season.

The 4-month breeding season is long enough for the birds not to be synchronized: some nests will have eggs that are being incubated or even eggs that have hatched at the same time that laying has just commenced in others. There is a range of possible starting dates. Table 5.9 examines the mean survival periods of nests started before or after the median nest starting date for all nests in each year. There was no clear effect of starting date on nest survival, although possibly early nests survived a little better. In addition, if a nest was destroyed during the laying period, the birds sometimes nested a second time (53% of 17 males did make a repeat attempt the same season), so that nesting early in the season gives a greater opportunity for a second nest that year if the first nest fails. There are, of course, other possible advantages and disadvantages of starting early or late, which depend on differential chick survival but no data are available.

6   Strategies Adopted by Major Females

Inspecting nest

The factors that determine whether a female ostrich embarks on a major or minor role, and the relative reproductive pay-offs of the two different strategies, are considered in Chapter 7. In this chapter, for major females already having begun a nest, I will consider the options open to them at various stages, and the reproductive consequences of the choices they make. I use and analyse the data in previous chapters and present further more detailed data.

6.1   Tolerance of Minor Females

Major female ostriches were remarkably tolerant of minor females coming to lay in their nests. As described earlier, a major hen would usually rise from her nest upon the arrival of a minor hen wishing to lay, and then wait nearby until the minor hen had finished. Given most animals' refusal to accept and care for offspring other than their own, this tolerance is surprising. I consider first what are a major female's options other than to be tolerant in this way, and then try to quantify the consequences of her tolerance.

It is likely that it would be exceedingly difficult, and possibly risky, for a major female to attempt to prevent another hen from laying in her nest, largely because of the vulnerability of her eggs to breakage and to

Plate 6.1   A major female moves a little away from her nest when a minor female comes to lay in it.

predation. Should she try, her possible courses of action are either to remain seated firmly on her nest or to chase away the minor female. Neither is likely to be successful.

One major female was observed to remain sitting at the nest while a minor female stood above her clearly wanting to lay. The minor female started after a while to peck, fairly gently, at the head of the sitting major female, and continued to do so intermittently for 20 min until the major hen eventually got up. The mere fact that the major female is sitting places her at a disadvantage to the minor female. One might expect the minor female to suffer a similar disadvantage when she in turn sits to lay; however, females are very quick to lay their eggs. (Fig. 6.1 shows the median length of time to be 50 s.) Thus there is an asymmetry which operates to the advantage of the newcomer.

A second asymmetry operates when considering the costs of physical competition at the nest, if actual fighting were to take place between the two females. Because the major female is the first to lay in her nest, at any stage she has more eggs in it than any minor female, and usually several more; thus, she would incur on average a greater cost if any eggs were broken in the course of any conflict. In addition, the breakage of any eggs would also be likely to render the whole nest more vulnerable to detection

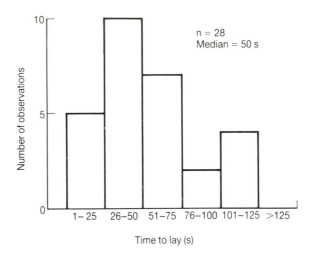

Figure 6.1   The time taken for female ostriches to lay their eggs, i.e. the time the bird sits at the nest. The data are for 1979 only, for major and minor females.

by hyaenas and jackals hunting by smell. Physical conflict between ostriches would also be conspicuous and likely to attract the attention of diurnal predators such as lions, which are often on the lookout for unusual behaviour on the part of potential prey animals. The risk of general injury to the combatants is significant, and in particular the breakage of an unlaid egg would be likely to have serious consequences.

A third asymmetry arises in this situation: the minor female has more at stake at that particular moment than the major female, and so is likely to compete harder. By depositing her egg in the nest, the minor female gains one more egg's worth of reproductive output; by preventing her, the major female would not gain as much.

There is in any case no reason to expect the major female to have any particular advantage in any conflict. No direct information was gathered on wild birds' weights, but there were no clear indications that the major hens were generally either larger than, or dominant over, the minor hens. For example, they did not in general lay larger eggs (Section 6.2), major females could become minor females and vice versa (Section 7.1), and there were no indications of a dominance hierarchy when away from the nest (Section 7.3).

To prevent minor hens from laying in her nest, would require a major hen to be in attendance – and therefore not feeding – for a greater proportion of her time than would otherwise be necessary. Figure 4.3 shows that particularly during the early part of the laying phase, the major females were often not present at their nests; this was especially the case on those days on which they themselves did not lay, but it was usually on those days that the minor females tended to lay eggs there (Section 4.3). The males were quite often present but could not necessarily be expected to try to prevent the minor females from laying. The males would have mated with these minor females also (Section 8.3), and would thus have had a genetic interest in the welfare of their eggs.

A major female, because she attended the nest for much of the time, could in theory later take sanctions against a minor female who laid an egg in her nest by destroying that egg. This would have the disadvantage of attracting predators to the smell, unless she were to eat the contents. One might expect this latter course to be an advantageous one for her to pursue, especially if augmenting her diet in this way enabled her to lay more eggs. (The factors influencing how many eggs female ostriches laid are considered in Section 6.3.) If this was a possibility it would make redundant the need for conflict with minor females about their laying. The fact that major females do not take this course of action suggests that there may be reproductive benefits to them from allowing minor females to lay in their nests. These possible benefits are now considered.

One possible way in which a major female's inclusive fitness might be increased by allowing minor hens to lay in her nest would be through kin selection (Hamilton, 1964), provided that those minor hens were closely related to her and provided that they would otherwise be unlikely to be able to find a suitable nest in which to lay. Figure 6.2 shows in diagrammatic form the periods in 1979 for which nests in the study area were available (i.e. were in the laying phase). It can be seen that there was usually a choice of nests available – although not always a wide choice as not all the nests would necessarily be known to a minor female at the time she wanted to lay. Indeed, it might not be until several days after a nest had been begun by a major female that minor females became aware of it, as was possibly suggested by the slower initial rate of growth of nests (Fig. 4.7). On the other hand, the choice would probably be considerably increased by edge effects; thus females could go outside the area in search of nests, although they might not be so good at finding them in regions where they had spent little time previously and where the males might be less willing to show them their nests.

Thus it is quite possible that relatives would indeed benefit from being allowed to lay in a major female's nest. The remaining question is whether minor hens were likely to be close relatives of the major hen. The ostrich social system, with promiscuous mating (Chapter 8), communal laying (Chapter 4), crèches for the young (Chapter 5) and high adult mobility and flexibility of companionships (Chapter 3), means that the average degree

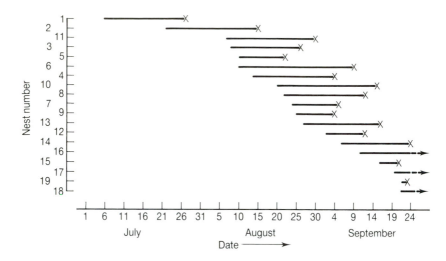

Figure 6.2 The availability of suitable nests for minor females to lay in. For each nest known in 1979, the range of dates the nest was in existence and still in the laying phase is indicated.

Table 6.1

The mean relatedness values among birds from a typical ostrich nest

Mean relatedness between any two chicks hatched in the same nest	0.21
Mean relatedness between a chick and the major hen who hatched it	0.28
Mean relatedness between a chick and the male in whose nest it hatched	0.35
Mean relatedness between a chick and a minor hen who laid in the nest in which it hatched	0.08

of relatedness between any two birds will be extremely low. Nor is there any very probable means by which an ostrich can recognize close relatives. It is likely that chicks are able to recognize the hen that hatched them, but whether they can retain that recognition despite being reared entirely by another hen is unknown. It is likely too that chicks are able to recognize other chicks that hatched in the same nest as themselves. Table 6.1 shows the average degree of relatedness among birds connected with a particular nest. The calculations are made assuming the following about a typical ostrich nest in Tsavo:

1. That 20 eggs are incubated in the nest (Table 4.3 shows a mean of 19 eggs per nest; 20 is used for convenience of calculation – it makes a negligible difference to the result).
2. That 11 of these were laid by the major hen (Table 4.3).
3. That all the major hen's eggs were fertilized by her mate.
4. That the male fertilized a random 33% of the nine incubating eggs laid by minor females (Section 8.3).
5. That these nine eggs were laid by three minor females who each laid three of them (Section 4.3).
6. That fertilization of the remaining 67% of these eggs was shared randomly among three other males.

It is clear that the average degree of relatedness between a chick and the major hen who hatched it, is fairly low (0.28). It is also clear that since a minor hen lays in more than one nest (Section 7.1), she can only be as closely related as this to the major hen at one of those nests, because mothers are unique. Her average degree of relatedness to those other major hens who are old enough to be her mother will be much lower; the highest would be 0.08 if they were minor hens who had laid in the nest in which she had hatched. Similarly, the average degree of relatedness among nest mates is low (0.21), and the pattern of laying in different nests (Section 7.1) renders it most unlikely that minor and major hens are usually hatched in the same nest. The conclusion is that kin selection is unlikely to be an important factor in the evolutionary maintenance of a major female ostrich's acceptance of other females' eggs in her nest. Nor was there any evidence of a reciprocal tendency for tolerance between the major and the few minor females, i.e. there was no indication of 'I let you lay my nest because you let me lay in yours'.

A second possibility is that a major female's own eggs benefited from the presence of other females' eggs. They could be protected against predation by being diluted in the same way as has been considered for chicks (Section 5.5). Dilution would protect a major hen's own eggs provided that nest destruction by predators was not always complete, and provided that larger nests did not suffer proportionately more successful predator attacks. There is some information on both points. Table 6.2, which gives data on partial nest destruction, considers only those 16 nests which survived to incubation and where it was known whether or not any breakages had occurred during the laying period. The causes of breakage were rarely known: some were certainly inflicted by predators and some were probably accidental. Table 6.2 shows that partial loss of eggs was frequent, even in nests which survived, averaging 1.13 eggs per nest, and that both major and minor hens' eggs were broken.

Table 6.2

Egg breakages in surviving nests

Nests which survived until incubation where it was known whether or not breakage had occurred during the laying period	16
Nests (included in the above) which suffered one or more breakages	8 (50%)
Eggs broken in these nests	18
Mean breakages per surviving nest	1.13
Eggs broken which were laid by:	
major hen	10
minor hen	4
unknown hen	4
Breakages which occurred	
nest size 1–10	7
nest size 11–20	4
nest size 21–30	4

Table 6.2 also indicates that larger nests were not proportionately more liable to suffer breakages than smaller nests. The size of the nest when breakage occurred could not always be accurately determined, but using a crude classification of small (1–10 eggs), medium (11–20) and large (21–30) nests, a proportionate increase in the number of breakages with the number of eggs in the nest was not seen. In contrast, experiments with dummy nests (Section 5.3) did show an increase in vulnerability to predation with nest size. The difference is presumably due to the fact that the ostriches increasingly attended and guarded nests as the latter grew older and therefore larger. Thus in real nests, the effects of age, size and guarding could not be separated. The higher proportion of major than of minor hens' eggs being broken was probably because they made up a greater proportion of the eggs in the early nests, in which many of the breakages occurred.

The degree of protection afforded by dilution of the eggs thus appeared to be slight. If we assume that a mean of 1.13 eggs was going to be broken randomly at each nest regardless of the number of eggs or of hens laying there, then the loss to a major hen in a nest containing only her own eggs would be 1.13, whereas the loss to a major hen who had laid 11 or 12 of the total of 26 eggs laid in her nest would be 0.52 eggs, i.e. $1.13 \times (12/26)$. Thus, on average, by diluting her eggs with others, she saved a maximum 0.6 eggs of her own. In addition, she may have benefited later to a more

Plate 6.2   An incubating nest seen from above. Note the warm central clutch being incubated, and the untended eggs around the outer edge of the by then large dusty nest area.

marked extent by having her own chicks diluted and thereby similarly provided with some protection against predation.

Protection by dilution was only worthwhile for the major hen if she had spare capacity, i.e. if she could sit on more eggs than she could lay herself, or if she could guard more chicks than she could hatch herself. Factors influencing the numbers of eggs laid and incubated are discussed below (Section 6.3).

6.2   Arrangement of Eggs

As described in Chapter 4, at most nests a number of eggs were pushed out of the central bowl when incubation began, and remained outside unincubated and doomed. Let us examine why this rearrangement was carried out, by whom, how and with what result.

Successful incubation requires the egg to be maintained at a fairly precise constant temperature for a considerable period of time (Drent, 1975). It was clear that in order to achieve this, the ostrich must cover completely all the eggs which it is attempting to incubate. It was also clear

Table 6.3

The ownership of eggs in the central group *vs* eggs pushed out

Nest		Laid by	No. of eggs in centre	No. of eggs in outer ring	P
A	1973 (nest A)	Major hen	9	0	0.02
		4 minor hens	10	8	
B	1978 (nest 1)	Major hen	13	0	0.01
		2 minor hens	6	5	
C	1978 (nest 4)	Major hen	9	1	0.03
		4+ minor hens	9	10	
D	1979 (nest 2)	Major hen	12	0	0.17
		3 minor hens	13	3	
E	1979 (nest 8)	Major hen	8	0	0.50
		2 minor hens	7	1	

that in the larger nests during the laying period, the birds were unable to do this – eggs were visible, protruding from under the feathers of the sitting bird. It would appear that about 20 eggs is the maximum number that can be completely covered by a sitting ostrich. There is ample evidence from farming practice with domestic poultry that a bird which attempts to incubate too large a clutch achieves an absolute lower hatching success than a bird incubating fewer eggs. The same almost certainly applies to ostriches. In the domesticated ostrich farming industry, birds are given 16–20 eggs to sit on, since this is considered to produce the best hatch (personal communication with several farmers). Thus wild ostriches were almost certainly pushing out surplus eggs, and retaining roughly the number required to result in the maximum number of chicks from that nest.

Time-lapse photography at four nests has shown that the pushing out of eggs from the central group to the outer ring was performed by the major females. We have 15 instances on film of the position of eggs changing from inside to outside the nest while the major female was the sole bird at the nest; in contrast, there were no cases of this happening while the male was at the nest, nor while minor females were present. Thus the arrangement of eggs was effected by the major female. It was spread over periods of 1–5 days in an apparently disorganized way. Eggs might be pushed out in any direction, and in the first day or two an egg which had been expelled might be returned to the central group of eggs.

Table 6.3 shows the ownership of eggs both in the centre and in the outer rings of five nests. Only five of 57 nests were chosen because they had to satisfy three criteria: (1) the nests had to survive the 3-week period before incubation started (only 22 nests did survive this long); (2) the nest had by then to contain enough eggs, i.e. more than 20, for some to be pushed out (only 17 of the 22 nests did); (3) the nests had to be discovered early enough to determine which eggs were laid by the major females and which eggs by the minor females (this was the case in only five nests). Seven nests were found so late that egg rearrangement had already taken place and incubation was under way; three others were at the hatching stage (Table 5.2).

Table 6.3 demonstrates that, with only one exception, every egg pushed out was a minor hen's egg. It can be seen, too, that this selective eviction of minor hens' eggs was most unlikely to have occurred by chance, and that four of the five major hens did as well as they could given the number of eggs pushed out. At nest D, it is likely that further eggs would also have been pushed out had not the nest been destroyed by predators a couple of days after incubation had started (as revealed by examination of a few surviving eggs). It is perhaps significant that the one major hen's egg that was pushed out, at nest C, was her first egg in this nest; as such, it would have been more exposed to heat stress than any other egg, and it would be less certain to have been fathered by the male than the major hen's other eggs. Thus if a mistake of this sort was to be made, it happened to the 'right' egg; however, there is no information on when or why this egg was evicted, nor by whom.

It is worth examining here the possible cues which major females might have used to enable them to discriminate between eggs in favour of their own.

a) *Position in the nest*. In principle, if the major hen always laid at one end of the nest and later pushed eggs out from the opposite end, she might be able to favour her own eggs. However, it can be shown that this was not the case. Figure 6.3 plots the positions of the major female's eggs in nest C and then again 5 days later during the pushing out of surplus eggs. At the earlier date, they were scattered throughout the nest rather than being clumped together. Similarly, the minor hens' eggs which would later be evicted were also scattered throughout the nest. All eggs were moved about in the nest to a considerable extent during the laying period, and Fig. 6.4 shows examples of such movements. Both major and minor females were seen (and occasionally even heard) moving eggs, particularly just after laying one. Where an egg ends up bears little if any relation to where it was laid. It is clear that an egg's position in the nest is not a sufficient means by which a major female can discriminate in favour of her own eggs.

(A) Before pushing out began. 29/8/1978

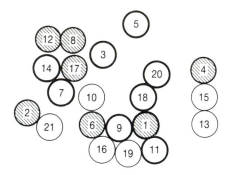

(B) During pushing out. 3/9/1978

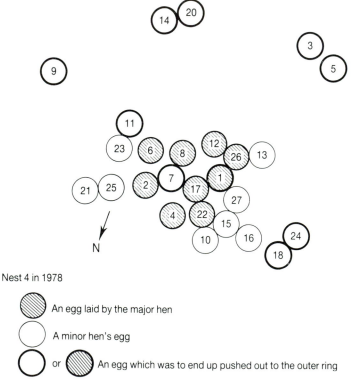

Nest 4 in 1978

An egg laid by the major hen

A minor hen's egg

or   An egg which was to end up pushed out to the outer ring

Figure 6.3   The positions of eggs in a nest before and during the pushing out of surplus eggs.

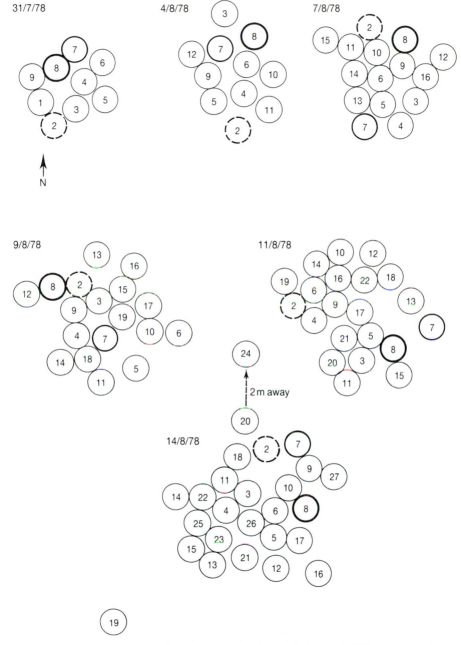

Figure 6.4   The movements of eggs in a nest during the laying period. The movements of eggs numbered 2, 7 and 8 are particularly indicated.

Plate 6.3   A female ostrich moves the eggs beneath her. There is ample scope for her to observe the eggs in her nest very closely.

b) *Age of eggs.* As was shown in Fig. 4.7, minor hens' eggs laid in a nest tended to be laid later than those of the major hen. Therefore, if the major hen were able to distinguish the age of the eggs in her nest and were to expel the most recently laid, she could achieve a distribution that would not necessarily be optimal but which would be better than random. Figure 6.5 shows, *inter alia*, the ages of the eggs pushed out of one nest. There is evidently no indication that the major female was using age as a cue, first because there was no age difference between the minor hens' eggs retained and those evicted, and second because she did not expel any of her own recently laid eggs. Information from the other four nests similarly showed no signs of major females distinguishing between eggs on the basis of their age.

c) *Size of eggs.* Individuals were reasonably consistent in the size of eggs they laid. All the eggs of the major female at nest C (Fig. 6.5) were smaller than all the eggs of the minor females and therefore she might have been able to use size as a cue to favour her own eggs. Major females would have to use not size *per se* but size in relation to the size of their own eggs, since as Fig. 6.6 shows, at nest B the major female's eggs were all larger than those of the minor females. At nest A, they were almost all smaller (one

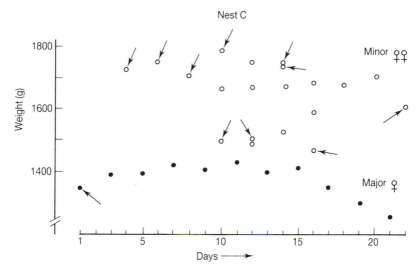

Figure 6.5   The laying date, weight and fate of eggs in nest C. The eggs arrowed ended up in the outer ring.

overlap only), but at nests D and E there was complete overlap. However, if the major female of, for example, nest C (Fig. 6.5) were using size as her sole means of discriminating, we could expect that she would have expelled the largest of the minor hens' eggs; however, this was not the case. As Table 6.4 shows, there was no appreciable difference in the size of those minor hens' eggs retained *vs* those expelled. If any bird was using size alone to discriminate between eggs, it would have to be able to detect minute differences in linear dimensions, as the weight difference was often only 35–40 g (2.5%).

d) *Shape of eggs.* Some individuals laid eggs that were fairly consistent in shape. The length:width ratio provides one measure of the shape of an egg. Figure 6.7 shows that successful discrimination by shape alone would have been theoretically possible in nest B where there was no overlap in the length : width ratios of the major and minor hens' eggs, but impossible in the other four nests. There was no relation between the size and the shape of eggs, either within or between birds. Among minor hens' eggs, there were no consistent differences in shape between those expelled and those retained. Some individuals laid eggs with unusually pointed ends, and a few laid eggs with a slightly swollen end, and in some cases these features might also have provided cues for successful discrimination.

e) *Surface appearance.* Close examination of the shell surface, especially

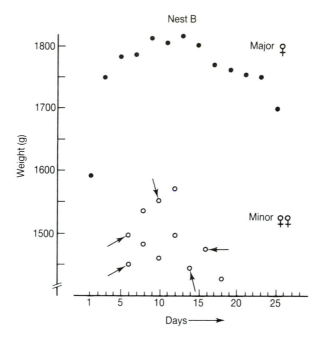

Figure 6.6   The laying date, weight and fate of eggs in nest B. The eggs arrowed ended up in the outer ring.

when an egg had dusty pores but an otherwise clean surface, sometimes revealed consistent qualitative differences in shell characteristics. For example, some had larger, deeper, narrower or more abundant pores than others; some were slightly whiter and some creamier in colour than others; and some had ends more devoid of pores than others. These characteristics

Table 6.4

Mean (±s.d.) of eggs at five ostrich nests

	Major hen's eggs (g)	Minor hens' eggs (g)	
Nest		Retained	Evicted
A	1378 ± 40	1556 ± 70	1596 ± 58
B	1748 ± 61	1487 ± 48	1476 ± 40
C	1361 ± 54	1628 ± 81	1641 ± 116
D	1672 ± 81	1646 ± 159	1609 ± 153
E	1612 ± 23	1540 ± 158	1270

Note: The numbers of eggs in each category are shown in Table 6.3.

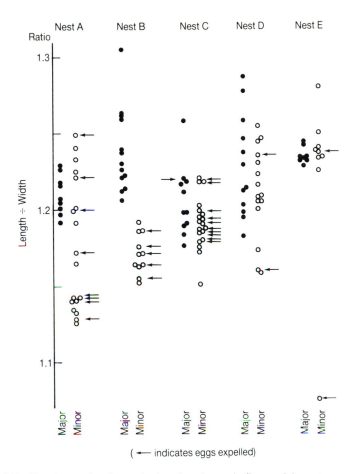

Figure 6.7   The shape of major and minor hens' eggs in five ostrich nests.

could provide cues for discrimination, although the differences were not easy to detect when the eggs were smeared and tinged with the red dust that was characteristic of Tsavo.

f) *Smell.* Discrimination by smell might have been possible. It was unlikely, however, for the following reasons. First, discrimination would presumably have been difficult between eggs that were up to 3 weeks old and which had each been subjected to similar influences in the nest. Second, egg odour is likely to be strongly selected against in ostriches because eggs that smell are going to attract nest predators. Third, ostriches, like most other birds, show little signs of response to olfactory cues in other aspects of their behaviour.

g) Observers' markings. It is highly unlikely that any of the ostriches could have made use of the numbers scratched onto the eggs for identification. These markings did not themselves contain clues as to which bird laid each egg.

Thus it appears probable that the major hens used a combination of surface appearance, shape and size to identify their own eggs. If such a combination was used, testing the importance of each becomes extremely difficult, compounded by the fact that it is possible that each major hen does not specifically recognize all her own eggs and therefore retain them, but may instead be able to distinguish which eggs are most alien. For example, a fibreglass dummy egg containing a temperature recording device was apparently detected as being most alien, and was the first of any eggs to be evicted. If the major hen were to find a particular discrimination difficult, she could presumably play safe and retain that egg, expelling only those which were more clearly not her own.

With only two exceptions (out of 15), every minor hen known to have laid more than two eggs in these five nests had some of her eggs retained inside the nest and one or more pushed out. Thus discrimination by the major hen was not effected on the basis of recognizing a particular minor hen's eggs and expelling them; rather, she appeared to recognize her own eggs and retain them.

Egg rearrangement thus conferred a considerable advantage on major hens, and it was possible to calculate how great and how general this advantage was. In the five nests in Table 6.3 for which information was complete, 98% (51 of 52) of major hens' eggs were incubated, whereas only 63% of 72 minor hens' eggs were incubated. However, this is a small sample, and it does not take into account those nests where no surplus eggs were pushed out, nor the fairly frequent very large nests which were not included in our small sample. Table 6.5 remedies this: it is an extension of Table 4.3 for all 20 incubating nests. For 11 of these nests, major and minor hens' eggs could not be distinguished with certainty. For these nests, we can extrapolate from the nine better known nests. Since the number of eggs laid by the major hens averaged 11 ± 1.9 eggs, and since it was shown in Section 4.4 that the number of major hen's eggs bore no relation to the total nest size, this average figure of 11 major hen's eggs has been used for these 11 less well-known nests (except in one nest where supplementary information was available). Therefore, it is reasonable to assume that on average the number of eggs laid by minor hens was the total in the nest minus 11. Virtually all (98% above) of major hens' eggs remained in the centre of the nest; therefore, the number of minor hens' eggs in the centre of a nest could be calculated by deducting 11 from the

Table 6.5

The numbers of major and minor hens' eggs incubated or evicted

		No. of eggs present			No. being incubated			
Year	Nest	Total	Laid by major hens	Laid by minor hens	Total	Laid by major hens	Laid by minor hens	No. evicted
1973	A	27	9	18	19	9	10	8
1977	2	20	11	9	20	11	9	0
	13	22	(11)	(11)	20	(11)	(9)	2
	14	36	(11)	(25)	22	(11)	(11)	14
	16	23	(11)	(12)	16	(11)	(5)	7
	17	36	(11)	(25)	21	(11)	(10)	15
1978	1	24	13	11	19	13	6	5
	4	29	10	19	18	9	9	11
	5	20	(4)	(16)	14	(4)	(10)	6
	6	34	(11)	(23)	21	(11)	(10)	13
	9	25	(11)	(14)	22	(11)	(11)	3
	11	20	(11)	(9)	18	(11)	(7)	2
	14	24	(11)	(13)	20	(11)	(9)	4
	15	31	(11)	(20)	18	(11)	(7)	13
1979	2	28	12	16	25	12	13	3
	4	15	11	4	15	11	4	0
	6	17	14	3	17	14	3	0
	8	16	8	8	15	8	7	1
	11	25	(11)	(14)	21	(11)	(10)	4
	13	21	10	11	21	10	11	0
Totals	n = 20	493	212	281	382	211	171	111

Notes: Table 6.5 is an extension of Table 4.3. Figures in parentheses are estimates (see text).

total number – the remainder being in the outer ring. The values obtained by extrapolation in this way are shown in parentheses in Table 6.5. The totals in Table 6.5 show that only about 61% (171 of 281) of minor hens' eggs were incubated; this figure differs little from that obtained from the smaller sample above. The major hen's ability to discriminate when rearranging eggs gave her a 58% advantage over the minor hens, such that just before incubation, her 11 eggs had a reproductive value equivalent to 17 or 18 minor hens' eggs.

Since the major hen was capable of discriminating between her eggs and the minor hens' eggs, it is worth considering why she did not expel all of the latter. Three possible reasons can be put forward:

1. The possibility of errors. Only one of 52 major hens' eggs was shown to have been expelled and therefore doomed. Even with only a 2% probability of making mistakes, to push out the average number (8.6 from Table 6.5) of minor hens' eggs also being incubated would mean on average the loss through similar mistakes of 0.17 major hens' eggs per nest. It is also possible that if the major female had already excluded the eggs which she found it easiest to discriminate against, the likelihood of mistakes would increase further with subsequent expulsions.

2. Dilution in the nest. As during the laying period, the dilution of the eggs in the nest through retention of other eggs (Section 6.1) may have helped to protect the major female's eggs during partial nest destruction. The selective advantage here was slight because egg loss due to predators from incubating nests was very low: if the nest was plundered, it was usually completely destroyed.

3. Dilution of chicks. As considered in Section 5.5, the protection gained by chicks who joined with others was probably appreciable. Retaining some minor hens' eggs in the nest provided group companions from a much earlier age (i.e. from the date of hatching) than was possible by joining broods (median age at first merger was 14 days in Nairobi National Park: Hurxthal, 1979).

6.3 Laying More Eggs

The reproductive output of a major hen from her own nest would clearly be greater if she were the mother of all 19 eggs being incubated in her nest. Why therefore did she not lay more eggs herself? Let us examine several factors which were likely to be involved.

a) *Physiological limitation.* In captivity, ostriches are essentially indeterminate layers (Smit, 1963), regularly producing up to about 40 eggs if these are removed as they are laid and if there is sufficient food. There have been cases of domesticated ostriches laying dozens of eggs in a year (Smit, 1963). Wild birds obviously do not lay as many as this; information on the numbers of eggs laid is given in Section 7.2. Attendance at the nest for much of the day reduced the major hen's feeding time and probably her food intake (Table 7.5). Nesting took place in the dry season when food was probably more scarce and when protein levels in the surrounding vegetation were probably at their lowest. There are no data on ostriches' ability to find, store or mobilize the $c.$ 300 g of calcium needed for each eggshell. Laying stopped roughly when incubation commenced, but cause and effect are not understood, and there may be other factors complicating the changeover (see below and Section 8.4).

Plate 6.4 A bisected (hard-boiled) ostrich egg, with a chicken's egg for comparison. Note the thickness of the shell, and the volumes of calcium, protein and fat required.

b) *Decline in hatchability*. If a major hen were to lay her whole clutch herself, the first egg she laid would be 38 days old on the day she laid her twentieth egg (assuming one broken and 19 surviving eggs in the average nest). For the eggs of domestic ostriches in artificial incubation conditions, Fig. 5.2 shows that there was a sharp decline in the viability of eggs with storage periods in excess of 15 days. Wild ostriches achieved higher hatching rates than those shown in Fig. 5.2. None the less, it is likely that the hatchability would be very poor for wild ostrich eggs that were 30–38 days old before incubation even commenced. How poor probably depends partly on the environmental conditions – Fig. 5.2 shows that the decline in hatchability over time occurred later with eggs that were stored in cool conditions.

c) *Risk of predation*. As shown in Sections 5.2 and 5.3, ostrich nests ran the perpetual risk of being destroyed by predators. The more eggs the major hen laid in her nest, the longer she delayed the start of incubation and thus prolonged the total period for which the nest was in existence and liable to be destroyed. There is an optimum number of eggs for achieving the probable maximum number of chicks from a nest: to lay fewer results in a smaller hatch, while to lay more results in a greater probability that the

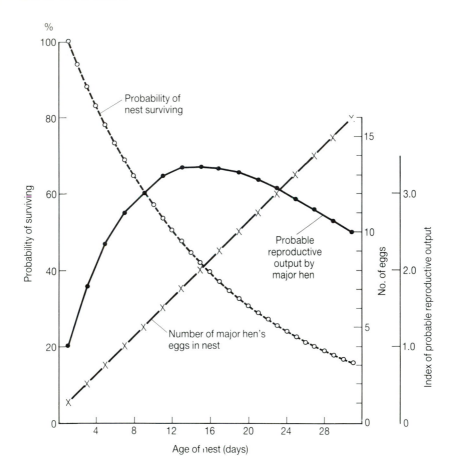

Figure 6.8   Changes in the survival probability of an average ostrich nest with time, and the probable reproductive output of the major hen there.

nest will be destroyed. What this optimum number of eggs will be depends on the length of the laying period and on the risks of predation during that period.

Data relating to these points have already been discussed. Assuming the mean daily risk during the laying period to be 0.06 per day (Section 5.3), and that the laying period increased by 2 days for each egg the major hen laid, Fig. 6.8 shows how the average major female's probable reproductive output from her own nest changed according to how long she went on laying. This is the product of the number of eggs in the nest and the probability of the nest's surviving until hatching. It can be seen that a peak was reached at day 15, the day on which she laid her eighth egg in that nest. The timing of this peak depends on how great is the risk of predation. As discussed in Section 5.3, this was very difficult to determine. It almost certainly varied from season to season and from one location to another.

With a daily destruction risk half of that assumed to be the average in Tsavo (i.e. 0.03 instead of 0.06), the optimum time for the major hen to start incubating would be day 31, when she had laid her sixteenth egg. On the other hand, if the daily risk of destruction was twice as great (i.e. 0.12), the optimum day to start incubating would be day 7, when she had laid her fourth egg.

Thus the risk of nest predation would have to be considerably lower than that prevailing in Tsavo for it to be worthwhile for a major female ostrich to lay all 19 eggs herself.

d) *Other birds*. As will be examined later (Section 8.4), the male may determine when incubation starts, and therefore he may control directly the number of eggs laid by the major female. The minor females may also have an influence.

One way in which the major female might be able to increase her reproductive output from her nest would be if she continued to lay even when incubation had commenced, provided of course that these late eggs could hatch at about the same time as the rest of the precocial brood. In two cases, the last egg laid by a major hen was probably just after embryonic development had begun in the other eggs in the nest, but not by more than 2 days. The accelerated development of eggs as a result of physical contact with more advanced eggs has been demonstrated in a number of species of birds (see Section 9.5), but two attempts to demonstrate it in ostriches have been inconclusive (Bertram, unpublished observations; R. Faust, pers. comm.). It is possible, but unproven, that some degree of communication takes place in synchronizing hatching among eggs which have been incubated for the same length of time.

7 Major and Minor Female Strategies

Showing nest site

I have described some of the major female's main choices and have outlined the reproductive consequences of her decisions at these points. As a result, it is becoming clear why, if she is a major female, an ostrich hen behaves as she does. We must now consider what a minor female's options are. I examine to what extent the minor hen strategy was consistent and whether there were different categories of minor hens linked with nesting stages. It is possible from the data available to quantify some aspects of the reproductive success of birds using either the major or the minor female strategies. Finally, I examine factors relating to the individuals themselves to see which were likely to influence the choice of strategy the bird adopted.

7.1 Who are Minor Females?

Minor female ostriches were defined as those birds laying in the nests of other females. As we shall see, although the distinction between major and minor was completely clear at any particular nest, the difference between the strategies of the two groups became more blurred over a longer time span.

Figure 7.1 shows in diagrammatic form which individuals laid how many eggs in which nests in the main study area in 1979. The major female at each nest is indicated; in each case, she laid more eggs in her own nest

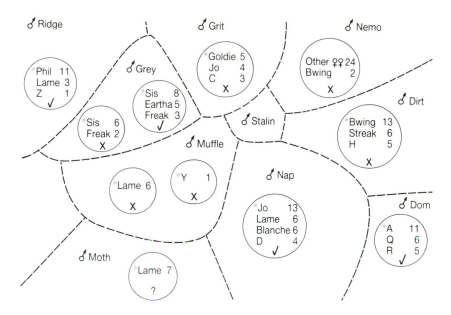

Figure 7.1   Which nests female ostriches laid in in 1979. The territories of different adult males are indicated by – – – –. Each nest is represented by a circle, within which is shown how many eggs were laid by each female. The major female at that nest is indicated by *. The fate of the nest is shown by √ = survived until incubation, x = destroyed before incubation, or ? = unknown.

than did any of the minor hens. Note that it is possible that there were other smaller, short-lived undiscovered nests in these males' territories, but it is unlikely that there were more than one or two. It is also likely that the females listed laid additional eggs in nests outside the intensive study area. Figure 7.1 illustrates that the minor females belonged to one of four distinct categories, as follows.

a) Pure minor females. Some individuals, such as Freak, laid eggs in one or more nests of other hens, and did not have a nest of her own. Freak was observed often enough for it to be certain that she did not play the role of a major hen elsewhere during the season; however, most of those birds classified as minor females could not be observed on a sufficiently regular basis. None the less, some of the other minor females, such as D, C and H, were unlikely to have been other than minor females during 1979.

The skewed sex ratio (Section 3.2), with some 40% more females than males in the adult population, meant that some females were inevitably without mates. Not all birds in this 'surplus' category, however, were necessarily without a mate throughout the season, because males whose

Plate 7.1   A major female at her nest; after it was destroyed she laid as a minor female in another bird's nest.

nests had been destroyed sometimes began to nest again later, usually with the same major hen but sometimes with a different one. An extremely high nest destruction rate and a large number of males switching partners between nests would be necessary for all hens to play the role of a major female in a single season.

We did not observe any case of a single female nesting alone, and neither has this been reported elsewhere. For a female nesting alone to be successful, she would have to go without food or water for a very long period of time, since leaving the nest unattended would greatly increase the risk of predation (Section 5.3). Similarly, no cases have been observed of two females only nesting together.

b) *Major females whose nests had been destroyed.* Female Jo laid eggs in Goldie's nest as a minor female after her own nest, where she had been the major female, had been destroyed by predators (see Fig. 7.1). Jo had stopped laying and had just started to incubate the eggs in her own nest when it was destroyed. Seventeen days later, she was laying eggs in Goldie's nest. In other cases, major hens have switched to a minor hen strategy with no break in their laying, as illustrated by Lame (see below).

c) *Future major females*. This category of minor females consisted of birds like Bwing who laid two eggs in a nest in male Nemo's territory and 4 weeks later started laying in her own nest as a major hen. Again, in other cases, birds have switched much more quickly, and in some cases without a break in their laying.

d) *Current majors*. This fourth category is very rare. There was only one case of a female with an extant nest laying an egg as a minor female in a different nest and then continuing to lay regularly in her own. From the regularity with which major hens' eggs usually appeared in their own nests, such switching was exceedingly unusual. We have no information as to what form of disturbance caused it in this instance.

Female Lame (see Fig. 7.1) illustrates beautifully the flexibility of the reproductive strategies. Early in the season, she was laying regularly as a minor hen in Jo's nest. She laid her last egg there just around the time that Jo started to incubate. (Unfortunately, Jo's nest was destroyed 2–3 days later.) There followed 8 days (and presumably four eggs) when it was not known where she was laying; circumstantial evidence suggests that she may have started as a major hen with male Muffle in her own nest which was destroyed early and never discovered by us. In any event, she did commence laying again as a major female at her own nest with male Muffle. This nest contained six eggs when it was destroyed by hyaenas. Lame immediately switched to female Phil's nest in male Ridge's territory; there she laid three eggs as a minor female, the first of these being laid on the day after Phil's last egg was laid and therefore around the time that incubation was beginning. After laying three eggs there, Lame started again as a major female, this time with male Moth. She had laid seven eggs there when the study period came to an end. Thus over a period of 50 days, she laid as a minor hen in different nests both before and after her own nests were destroyed, and she was twice a major female, with different males. Thus while it is meaningful to refer to major and minor hens *at a particular nest*, birds do change rapidly between major and minor strategies within a season, and therefore caution is needed when referring to major *vs* minor females.

It is in principle possible to estimate what proportion of minor hens' eggs laid in an area were laid by birds who had already been, or who would later be, major females. However, there are considerable practical problems in doing so, because some minor females could well have been major females outside the study area. Subjectively, perhaps 60% of minor hens' eggs were laid by birds who did not take up the role of major female at any time that season, and of the remaining 40% at least half were laid by birds who had been major females but whose nests had been destroyed. This last category in particular would depend on the prevailing rate of predation.

7.2   Performance of Major and Minor Females

It is useful to compare the reproductive performance of the major and minor strategies, but the interchangeability of birds between strategies makes this difficult. None the less, an attempt is made to quantify the numbers, sizes and chances of survival of both major and minor hens' eggs.

a) Numbers of eggs. There were only a few birds for which the total number of eggs laid in a season was known reliably, because they could have laid in other nests outside the intensive study area. A few birds remained within the study area, and data for the six females doing so in 1979 are given in Table 7.1. It can be seen that there was a considerable range in the number of eggs laid, between five and at least 22. The mean of > 14 eggs per female is a minimum figure because (1) other eggs not noted may have been laid by these birds during the study period; (2) one bird (Lame) was still laying on a regular basis at the end of the study period, and another (Jo) may also have been; and (3) Lame was strongly believed to have laid another four eggs in the middle of her laying bout.

   The data are somewhat scanty because of the difficulty of finding and observing individual females. Further information on the numbers of eggs laid was obtained by closely observing the less wide-ranging males. Table 7.2 shows the numbers of eggs in 1978 and 1979 known to have been laid in the territories of all males observed closely enough to be confident that few if any eggs were missed. Females were still laying in the territories of two males at the end of the 1979 study period, and so the mean figure for that year could be a considerable underestimate. The mean of 28 and 18 eggs per male corresponded to 20 and 13 eggs per female by correcting for the skewed adult sex ratio of one male to 1.4 females (Section 3.2). These minimum figures are of the same order of magnitude as those obtained for the six females above. The 1978 figure was higher than that in 1979, probably because of the higher rate of nest destruction in 1978; this results in fewer cases of major females stopping laying because of the onset of incubation.

   Among the six birds in Table 7.1, the fewest eggs (five) were laid by Freak, a minor female throughout the season; the next two smallest

Table 7.1

The laying records of six females in 1979

Female	No. of eggs laid	Laying history	No. of eggs still surviving at end of study period
Freak	5	As minor hen, only in Sis' two nests	2 (in 1 nest)
Phil	11	As major hen only, in her own early nest which survived	11 (in 1 nest)
Sis	14	As major hen only; her first nest was abandoned after a breakage half way through the laying period; her second nest survived	8 (in 1 nest)
Bwing	15	She laid two eggs as minor hen elsewhere long before starting as major hen in her own nest which was destroyed just before incubation started	0
Jo	? >17	As major hen in her own nest which was destroyed at the start of incubation; later as minor hen in another's nest, also destroyed; thereafter not known	0
Lame	>22	Minor–major–minor–major (see text)	>9 (in 2 nests)
Mean	>14		

numbers of eggs (11 and 14) were laid by Phil and Sis, who were both major females only; and the three largest numbers of eggs (15, 17 and 22+) were laid by Bwing, Jo and Lame, birds who used both strategies. For Lame and Jo it can be stated with confidence that they would not have laid as many eggs in 1979 if the nests in which they were major hens had not been destroyed. In general, laying in other birds' nests would appear to enable a female ostrich to lay a few more eggs than she otherwise would. Thus in some respects, the laying of a large number of eggs reflects not success but a high loss of eggs. Table 7.1 also shows the number of each female's eggs still surviving at the end of the study period. Due to the vagaries of predation, two of the three more 'successful' birds in 1979 happened to be birds that restricted themselves to a major hen strategy.

Table 7.2

Number of eggs laid in the territories of several closely observed males in 1978 and 1979

	1978		1979	
Male	No. of nests	No. of eggs	No. of nests	No. of eggs
Adolf	1	25	?	?
Dirt	1	31	1 (+1 other egg)	25
Dom	2	32	1	22
Grey	1	16	2	24
Grit	?	?	1	12
Muffle	1	4	2	7[a]
Nap	1	24	1	29
Nemo	?	?	2	28[a]
North	2	62	?	?
Ridge	1	31	1	15
Stalin	?	?	0	0
Mean		28		18

[a] Laying was still taking place in that territory at the end of the study period that year.

Whether or not a nest was destroyed by predators probably depended on chance, and a different set of random factors could equally easily have favoured different individuals. At the end of the study period, female Lame (Table 7.1) did not have all her eggs in one nest, and clearly had a greater probability of having an offspring hatch that year. In addition, despite the small sample, it is clear that laying in other birds' nests results in a female ostrich laying more eggs than she otherwise would if she laid in her own nest only.

b) *Sizes of eggs.* A total of 207 ostrich eggs were weighed in Tsavo West National Park in 1979, weighing between 1092 and 1970 g. As shown in Section 6.2, the eggs laid by individual females were reasonably consistent in size. One might expect that there would be a trade-off between the size of a female's eggs and the number she laid. Figure 7.2 shows that there was no such negative relationship among the six individuals in Table 7.1. Lame, who laid the largest eggs, laid more than did any other female; in contrast, Freak, who laid the smallest eggs, laid the least number of eggs. It did not appear that birds were able to lay more eggs by laying smaller ones. Rather, it appeared that the 'best' birds could lay both many and large eggs.

The assumption is that larger eggs are 'better' then smaller eggs, but

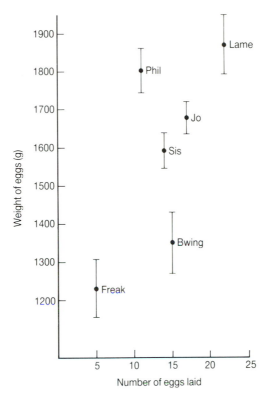

Figure 7.2   The relationship between the size of the eggs laid by female ostriches and the number of eggs they laid. Mean and s.d. is shown for each bird.

there are no data to prove this. After hatching, the chicks are vulnerable to predation largely because of their small size, and it is probable that the smaller chicks – from the smaller eggs – are more vulnerable than the larger chicks. Larger eggs are less vulnerable than smaller ones to over-heating if left unattended. On the other hand, it may be that larger eggs are more vulnerable to breakage in the nest because of reduced shell strength due to their reduced convexity, unless they have a correspondingly greater shell thickness. The relationship between egg size and shell thickness was investigated indirectly using the eggs of domestic ostriches at Oudtshoorn. A sample of 20 empty shells were weighed, and compared with the weights of 20 fresh eggs whose long and short cirumferences were both within 0.6% of those of the empty shells. (The cost of buying and emptying fresh eggs would have been prohibitive.) Among the 20 shells, there was as expected a significant positive correlation ($r_s = +0.71$,

Table 7.3

Mean (±s.d.) weights of major and minor hens' eggs

	Nest	Major hens' eggs		Minor hens' eggs	
		No. weighed	Mean wt (g)	No. weighed	Mean wt (g)
1977	2	11	1595 ± 51	11	1515 ± 94
	3	5	1384 ± 62	3	1557 ± 14
	8	3	1807 ± 39	5	1534 ± 101
1978	1	13	1758 ± 61	11	1492 ± 45
	4	11	1373 ± 52	19	1645 ± 101
1979	2	12	1672 ± 46	16	1639 ± 158
	4	11	1803 ± 59	4	1820 ± 160
	5	6	1569 ± 58	2	1154 ± 62
	6	15	1758 ± 55	3	1791 ± 20
	8	8	1612 ± 23	8	1507 ± 173
	10	12	1379 ± 47	7	1699 ± 85
	12	5	1512 ± 22	6	1672 ± 46
	13	11	1682 ± 31	11	1546 ± 119

Total 13 nests, in which major hens' eggs were larger in 7 and smaller in 6.

$P < 0.01$) between shell weight and the comparable whole egg weight. There was no significant correlation ($r_s = -0.23$, $P > 0.05$) between whole egg weight and the ratio of shell weight to whole egg weight. Thus larger eggs must have had slightly thicker shells. There are no data available relating to the risk of breakage of eggs of different sizes and with different thicknesses of shell, though it must depend on what object the egg is being bumped against.

There was no obvious relationship between the average sizes of major and minor hens' eggs in a nest. Table 7.3 shows that for the 13 nests for which there was enough information, the minor females' eggs were as likely to be larger as to be smaller than the nest owner's eggs.

It is worth noting that the eggs laid by most individuals showed consistent changes in size with laying order (Figs 6.5 and 6.6). The first two or three eggs laid, and the last one or two, tended to be smaller than those laid in the middle of the laying bout. The trend was more clearly visible with major hens, for whom there were fuller data, than for minor females, although within a single nest the latter often showed the same changes in egg size. The functional significance, if any, of the change in size is not clear.

Table 7.4

The chances of survival of major and minor hens' eggs

	Major hens' eggs	Minor hens' eggs	Unknown
A. *All 53 nests whose fate was known*			
No. of eggs surviving to incubation	234 (54.4%)	312 (61.7%)	0
No. of eggs destroyed before incubation	196 (45.6%)	194 (38.3%)	20
% Advantage of minor hens' eggs	13.4%		
Total	430	506	20
B. *The 37 nests discovered during the laying period*			
No. of eggs surviving to incubation	113 (38.8%)	133 (44.8%)	0
No. of eggs destroyed before incubation	178 (61.2%)	164 (55.2%)	13
% Advantage of minor hens' eggs	15.5%		
Total	291	297	13

Four (small) nests which were still surviving and being laid in at the end of the study periods have been excluded since their fate was unknown.

c) *Survival of eggs.* Figure 4.7 shows that the minor hens' eggs were not laid at a constant rate as was the case with the major hens' eggs, but were deposited during the second half of the latter's laying phase in particular. Nest predation, on the other hand, destroyed nests at all stages, as is shown in Table 5.2. Of the 29 nests destroyed within 22 days, ten (34.5%) were destroyed within half that time. Therefore, on average, the major hens' eggs ran a greater risk of predation or destruction by other means before incubation, because they had been laid earlier in the laying period than the minor hens' eggs. Table 7.4 shows the chances of both major and minor hens' eggs surviving until the start of the incubation period. Considering all those nests whose fate was known (Table 7.4A), it can be seen that 61.7% of minor hens' eggs *vs* 54.4% of major hens' eggs survived until incubation, giving minor hens' eggs a 13.4% advantage (61.7 − 54.4 ÷ 54.4). The sample was a large one but was biassed in favour of those nests which survived, because several nests were discovered only during the incubation period by observing birds approaching them. Therefore, the chances of survival of eggs in those nests discovered during their laying periods are shown in Table 7.4B; although the overall survival probabilities were lower, the differential between minor and major hens' eggs is similar (15.5%) to that in the larger sample.

Thus by laying in the later nests in particular, minor hens were able to improve the survival prospects of their eggs relative to those of major hens. On the other hand, as shown in Section 6.2, by her final arrangement of

Plate 7.2 A female on a nest may have to tolerate significant heat stress in the hottest part of the day.

the eggs in her nest, a major female was able to achieve a 58% relative advantage for her own eggs, dwarfing the 13.4–15.5% pre-incubation advantage of the minor hens' eggs. Overall, the eggs of major females had a considerably higher chance of being incubated, despite their smaller chance of surviving until incubation.

It is difficult to assess the costs of starting a nest, i.e. of becoming a major female. The energy costs involved in sitting on a nest are presumably less than those involved in standing and walking. Even when away from the nest, major females walked and ran significantly less often than other females (Table 7.5) by a factor of 1.8. On the other hand, they did not feed at a significantly different rate when away from the nest (Table 7.5) and since they spent an increasing proportion of their time attending the nest, their total food intake was presumably progressively reduced in either quantity or quality. No data relating to birds' weights or weight changes could be collected.

The mating rates of major hens were not significantly higher than those for other birds (Table 7.5), and all of the eggs were likely to be fertile anyway.

No case of predation on nesting ostriches was recorded, although such

Table 7.5

Activity of major and minor females

	Major females with current nests	All other females	Significance
No. of hours of observation of females not on nests	60.8	78.6	—
No. of observation periods	9	20	—
No. of different individuals observed	4	10	—
Mean % of time walking or running	18.6	33.3	$P < 0.01$
Mean distance travelled per 30 min (m)	300	308	N.S.
Mean % of time feeding or nibbling	35.0	39.6	N.S.
Mean mating rate per 10 h	2.53	1.71	N.S. ($P = 0.07$)

instances have been reported elsewhere. Sitting males, who attend the nests at night, are probably more vulnerable than females. It is difficult to envisage a predator successfully stalking a sitting female ostrich during daylight hours, for she has good all-round vision when her neck is up. The risk of predation run by minor female ostriches moving around the landscape is probably as great as that of a sitting major female; but in view of the birds' probable longevity, both risks are presumably very small.

7.3   The Decision to be a Major or a Minor Female

Understanding which birds become major females is confused by the adoption of different strategies by the same individuals. Thus most major hens were also briefly minor hens in any season, and some minor hens had been, or would later be, major hens. But not all of them became major hens, and it is worth loking at why some did and why some did not.

The birds' size might be a determining factor. The body size of wild ostriches could only be vaguely assessed. There was no detectable effect of size on a bird's likelihood of becoming a major female. One consistently major female, Bwing, was unusually small, whereas Jo was a little larger than average. The sizes of the eggs laid by the major and minor hens was loosely corrrelated with visual body size estimates, such that the smallest birds tended to lay the smallest eggs. Egg sizes could be accurately determined. However, as is shown in Table 7.3, there was no perceptible difference between the average sizes of the eggs of the major and minor females. Age may also have been a determining factor. There was no way

Table 7.6

The extent to which males paired with the same major females within and between seasons

	Same major female	Different major female
In a subsequent season		
No. of males who nested with	3	5
In a second nesting attempt within the same season		
No. of males who nested with	4	1

of measuring birds' ages, and therefore no way of determining whether older females were more likely than younger ones to become major hens.

No dominance hierarchy was apparent; the groupings of birds were so loose, and aggressive intereactions between the females generally so mild and infrequent, that a rigid linear hierarchy was unlikely. However, in 12 out of 13 (92%) chases, displacements and threats, where the status of the females was known, the chasing or displacing birds were major females. This does not necessarily mean that the dominant birds were major females simply because they were dominant; it may be that acquiring major female status confers dominance or increases aggressiveness. One particular female, Bwing, who was known to be a major hen in all three years, was not only unusally small but had an intermittently dislocating left wing which, to a slight extent, hampered her movements. It was most unlikely that she could have achieved dominance over large normal females, and indeed she provided the single instance of a major female being displaced by a minor female.

It was difficult to recognize more than a very few individual females reliably enough to be confident of their identity from year to year, because it was so difficult to observe them at close quarters; therefore, direct information on the consistency of female strategies is very sparse. However, similar information can be obtained from the males. There were eight cases where the identity of a particular male's major hen was known in consecutive years. In only three (38%) of these cases was she the same individual in successive breeding seasons (Table 7.6); in most instances, the 'discarded' major female was known to be still alive.

When a major hen's nest was destroyed by predators, she sometimes began another nest, again as the major hen. As shown in Table 7.6, this was more likely to be with the same male as on her previous attempt, but not invariably so.

It appeared that the process of becoming established as a major female was highly variable. In 1979, female Lame for example (Fig. 7.1), had become male Moth's major hen within only 8 days of the destruction of her previous nest with male Muffle; during that time, although she was laying as a minor hen in male Ridge's nest, she may have been spending most of her days with Moth. Similarly, she had earlier spent most of her time with her future mate Muffle while laying steadily as a minor hen in male Nap's nest. Also in 1979, female Bwing was clearly destined to become male Dirt's major hen at least 4 weeks before she actually started a nest with him. The two birds were frequently seen together, and if they were with other females, the male was likely to be closer to Bwing than to any other female. Bwing had been Dirt's major female in 1978; she was known not to have been in 1977 (when she was Nap's major hen), although she did lay three eggs as a minor hen in Dirt's nest that year.

Some females appeared to be more likely to become major females than others. Having been a male's major female before may to some extent have predisposed a female towards becoming his major female again the following year, but shorter-term factors were apparently in operation. The availability of females ready to start nesting (Chapter 8) and the destruction of nests (Chapter 5) probably also influenced who ended up as whose major hen.

8   Male Strategies

Mating

During the breeding season, male ostriches presumably have the option of whether to start their own nests. If a male does start a nest and it survives, he goes out of breeding condition at incubation, whereas if he does not have a nest, he remains in breeding condition for longer and continues to mate promiscuously with females who lay in others' nests. In this chapter, I consider first what proportion of males started nests, and at what stage of the nesting season. In later sections, I consider the costs and reproductive benefits of starting a nest, and examine possible areas of conflict between males and their major females.

8.1   Whether and When to Start a Nest

In the 1977 and 1978 seasons, no adult male in breeding condition was observed which did not have a nest. In 1979, on the other hand, three such males were noted. None of these was an individual known from previous years, and it is therefore possible that they were relatively young, maybe 3 or 4 years of age. One of the non-nesters, Fagan, was not territorial but was seen at different times in the territories of three different males. The other two non-nesters both held territories – that held by Stalin in the intensive study area is shown in Fig. 3.3. The other non-nesting male, Nicholas, held a similarly small territory, about 20% of the size of the

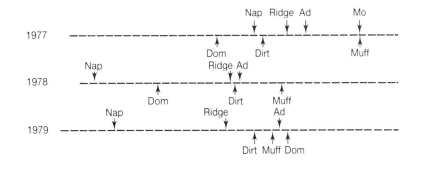

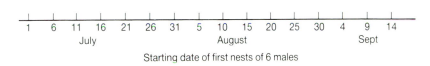

Figure 8.1   The starting dates of the first nests of six well-known males in each breeding season, 1977–79.

nesting males' territories. Thus we may tentatively conclude that these three non-nesters were probably not fully reproductive adults. The option of not starting a nest did not appear to be taken up by mature male ostriches.

Figure 5.6 showed that some males began nests as much as 3 months earlier than other males. Figure 8.1 shows the dates on which six well-known males which nested in each of the study years started what were almost certainly their first nests each year. For these few males, it was possible to assess the consistency with which individual males were early or late starters. There were positive correlations between years in the nesting order of these six males, but the effect was not a marked one. The correlation was significant between 1977 and 1978 ($r_s = +0.886$, $P < 0.05$), but not between 1978 and 1979 ($r_s = +0.371$, N.S.).

In Section 5.6, it was shown that nesting success was little influenced, if at all, by the stage of the breeding season. Thus there is nothing to suggest that nests started early result in improved chances of success. On the other hand, males whose first nests were begun early had more opportunity to make a repeat attempt if the first nest was destroyed. There was no correlation between the date of starting nesting and the size, location or vegetation type of the male's territory. Thus early *vs* late starting did not appear to be related to dominance or to any other measured characteristic of males or their territories; nor did it appear to result in any difference in the likelihood of progeny from that nest.

8.2   Costs of Starting a Nest

To some extent, it is possible to assess the energy and nutritional costs to males of starting a nest. During the continuous watches on known individual males in 1979, information was collected on the distance travelled during 30-min periods, and the incidence of feeding and movement recorded at 5-min intervals. This information has been arranged according to the stage of nesting of the male concerned. For nesting birds, there was no change in either of the two measures of movement with the stage of the nest (Fig. 8.2). The birds who did not have a nest at all that season travelled the shortest distances, but as described in Section 8.1, these non-nesting males had very small territories. Males whose nests had been destroyed also travelled less than nesting or pre-nesting birds. However, as is apparent from Fig. 8.2, there was considerable variation between birds and between the same bird on different days, and because the number of observation periods was small, the differences in the mean distances travelled are not significant. Figure 8.2 also shows that there were no significant changes in the feeding rates of males as their nesting advanced, although the mean feeding rate of nesting males was slightly lower than that of males who, for whatever reason, were currently without nests.

The data in Fig. 8.2 refer to locomotion and feeding only when not attending the nest. However, the males spent more of their time at their nests as the nests grew in size. Figure 8.3 shows these changes with nest growth. The data are derived from the time-lapse photography of the nests. At dawn, at dusk, and at each hour on the hour from 07:00 to 18:00, those birds present at their nests were recorded. It can be seen that with time, the males spent an increasing proportion of the daylight hours at their nests, reaching about 50% when incubation began.

Figure 8.4 shows the changes in the total daily amounts of locomotion and feeding during nesting. These are derived from the data in Fig. 8.2 on locomotion and feeding rates when away from the nest, corrected for the mean amounts of time each day that a male was away from the nest during the different nesting stages. Figure 8.4 shows that nesting involved the males in no extra total locomotory effort (because the greater rate of activity was outweighed by the reduced time in which to be active), and therefore there is probably little if any extra energy cost of nesting. However, since nesting males when away from their nests did not

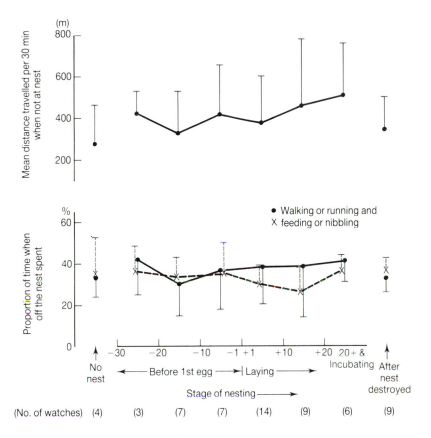

Figure 8.2 The activity rates of males at different stages of nesting, when not at the nest, in 1979. Means and s.D. are indicated.

apparently consume food at a noticeably faster rate than males without nests, they achieved a total daily food intake lower (by some 45%) than those males without nests. There must therefore be a nutritional cost of nesting. It is not known whether the shorter feeding time reduces only the quantity of food ingested, or whether the birds feed any less selectively and therefore experience any drop in food quality in their efforts to maintain the same quantitative intake.

Nesting males may be at a greater risk of predation as they attend the nest at night. All ostriches sit down at night whether or not they have nests, and may be found by a hunting predator. If there is any smell associated with the nest, the ostrich sitting there will be detectable at a somewhat greater range than one sitting roosting elsewhere. It is possible, too, that an ostrich sitting on a nest at night is both unable and unwilling

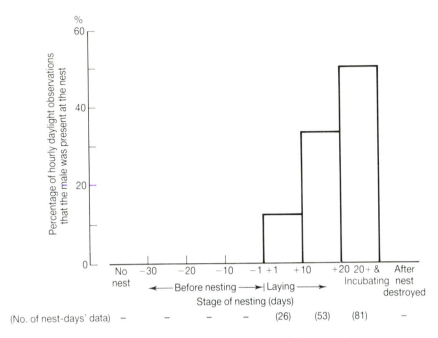

Figure 8.3 The proportions of observations during daylight hours that males were present at their nests.

to get up and run off as quickly in response to danger because to do so may well contribute to the destruction of the nest. Certainly, cases have been reported of male ostriches being killed by lions at their nests at night, but there are no quantitative data on the magnitude of the risk.

8.3   Benefits of Starting a Nest

Continuous observation of known individual males provided information on the numbers of matings by them within known periods of time. Figure 8.5 shows the mean daily mating rates of males, corrected for the time spent at the nest, at different stages of nesting. It is striking that the mating rates during a period 10 days before the first egg until 20 days after it, were at least eight times as high as before or after this period. Thus preparing for, and owning, an early-stage nest was correlated with a large but short-lived increase in the number of matings that a male achieved during that period.

The observation of males at the different stages of nesting was necessarily opportunistic, depending on when nests were found, and on when

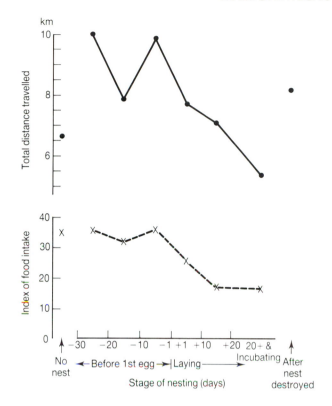

Figure 8.4   The total daily locomotion and feeding rates of males at different stages of nesting.

and whether they were destroyed. Many were repeat observations of the same individuals over a long period, whereas some males contributed information for only a short time. The picture could thus be somewhat confused by a possible seasonal bias, if matings were more likely to occur at a particular time in the breeding season as a whole rather than at a particular stage of an individual bird's nesting. Figure 8.6 shows the mean mating rates during 10-day blocks throughout the 1979 breeding season. Although there was a general rise in the mean mating rate, it was less marked than the rise and fall within an individual male's nesting phase. It correlates roughly with the dates (shown in Fig. 5.6) when new nests were being started.

The mean mating rate of males without nests is also shown in Fig. 8.5; this was far lower than that of birds who had, or were just about to have, an early-stage nest, but it was higher than that of birds with incubating

Plate 8.1 A male ostrich on his concealed nest suddenly becomes conspicuous, and he and his nest more vulnerable to predation, when a fire removes all the vegetation cover. The outer ring of doomed eggs is clearly visible.

nests. The male birds without nests belonged to one of three categories: (1) males who did not have a nest at all that season; (2) males who were to have a nest more than 30 days later; and (3) males who had had their nest destroyed and who had not started another by the time the study period ended. It is possible that some of this last category did start another nest soon after our departure, and that they should therefore have been allocated to one of the pre-nesting categories. In general, however, it can be seen that by starting a nest, a male achieved a considerable but short-lived increase in his mating rate, after which (if his nest survived) he had less frequent matings than males without nests.

Why did males with (or soon to have) early nests achieve higher mating rates? One reason is that they were likely to be accompanied by more females at that stage. Figure 8.7 shows, for different stages of the males' nesting in 1979, the mean number of adult females (recorded at 30-min intervals) who were to be found within 150 m of the males observed. It shows that a male had most female companions during the 10 days before the start of his nest and that there was a steady decline thereafter. This decline was partly due to the fact that as the nest grew, the major female was progressively more likely to be attending the nest, and therefore

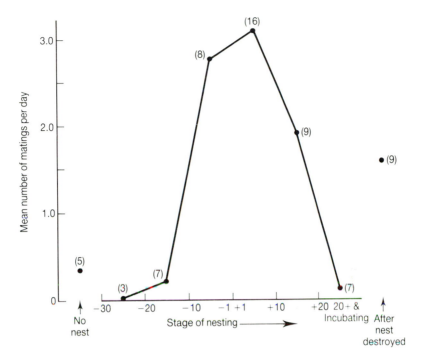

Figure 8.5   The mating rates of males at different stages of nesting in 1979.

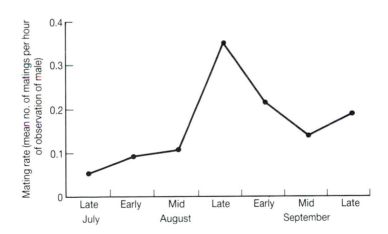

Figure 8.6   The mating rates of males at different times in the 1979 breeding season.

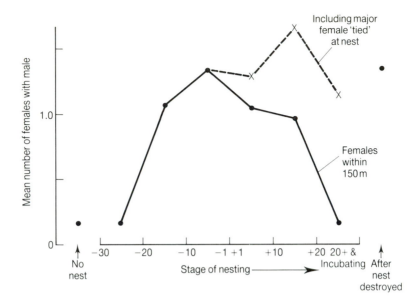

Figure 8.7   The numbers of females in the company of males at different stages of nesting.

unable to accompany him. However, she was in a sense 'tied' to the male in that she was in his territory, and was likely to rejoin him if she left the nest. Figure 8.7 also shows the male's mean number of companions if corrected for the probability (from Section 4.2) that his major female was at the nest and was in effect an extra 'tied' companion.

The rise and fall in the numbers of female companions with the stage of nesting was not due to a seasonal change in the number of females or in their gregariousness. Figure 8.8 shows the mean number of female companions, arranged into 10-day blocks throughout the breeding season; there was no change in female companionship with time.

Other possible causes of the increase in a male's mating rate included an increase in his courtship displays to the females accompanying him, or a change in those females' receptivity. The first of these cannot be confirmed, because there is little information regarding males giving the 'kantle' courtship display (see Section 4.1) to accompanying females. However, Table 8.1 shows the likelihood of their kantle displays resulting in copulation. Overall, 76% of kantle displays by males who either had, or were about to have, early nests (i.e. days -10 to $+20$) were successful, compared with a 59% success rate for displays by males who started their nests before or after this period. The difference was not significant.

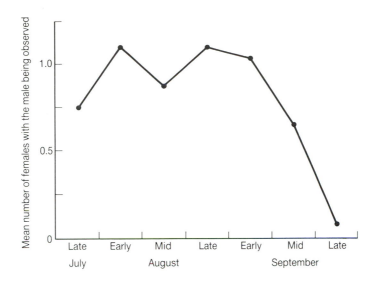

Figure 8.8   The numbers of females in the company of males at different times in the 1979 breeding season.

Table 8.1

The success of male's kantle displays in relation to their stage of nesting

	Copulation followed male's kantle display	No copulation took place
Male had or was about to have an early-stage nest (i.e. days −10 to +20)	28	9
Male did not have an extant early-stage nest, nor was he about to start one	16	11

$\chi^2 = 1.27$, N.S.

In assessing the reproductive benefits to a male of starting a nest, his mating rate alone is not a reliable absolute measure of his reproductive success for three reasons:

1. The number of matings per egg laid can vary.
2. A female may have mated with other males.
3. Some eggs are more likely to survive than others.

Table 8.2

The mating rates of major females compared with other females

	Major females	Known minor females	Females of unknown status	All non-major females
No. of observation periods	9	5	15	20
Mating rate (mean no. of matings per 10 h)	2.53	0.89	1.98	1.71

1. Table 8.2 shows the mating rates of known major females compared with known minor females and females of unknown status. It is likely that the great majority in this last category were in fact minor females, in that they probably neither had extant nests nor were about to start one. For this analysis, a female was classified as a major female if she had a surviving nest or if she was within 3 days of starting one. It is evident that there was considerable variation in the mating rates of female ostriches, and that there were only slight and non-significant differences in the mating rates of major and other females. It is clear, however, that on average females mated two to five times for each egg they laid. Although it was almost impossible to be certain that a female was not laying during any particular 2-day period, the indications were that females copulated for at least several days before starting to lay. The time relations between mating, the resultant fertilization of an egg and the laying of that egg are not known for ostriches. By comparison with domestic chickens, fertilization probably takes place about 3 days before the egg is laid, but there are no data on when the sperm which achieves that fertilization is likely to have been supplied.

2. Females mated with more than one male, and therefore the certainty of a male's paternity of eggs laid by a female with whom he had mated could be low. Table 8.3 indicates the categories of the two participants where known for matings observed during 1979, and shows that whereas a current major female was unlikely to mate with a male other than her present mate, minor females usually did so. Only a third of a male's matings with minor females were with birds currently laying in his nest, and assuming that those females were currently laying elsewhere, a male

Table 8.3

Numbers of matings by different categories of ostriches observed in 1979

Females	Males with extant nests or within 3 days before starting nests	Males without nests
Major		
Currently major at male's own nest	24	—
Currently major at another male's nest	0	2
Not major		
Currently known to be laying as minor in male's own nest	12	—
Currently known to be laying as minor in another male's nest	3	4
Female who did not lay in male's own nest and who was not known to be laying currently	5	1
Female of unknown status	2	9
Male's own future major, not currently known to be laying	—	2

92% (24/26) of matings by major hens were with their mates; 32% (12/38) of matings by minor hens were with the male whose nest they were currently laying in.

in general had only a 32% probability of paternity of the eggs that minor females laid in his nest. Three instances were observed of a female mating with two different males within a 24-h period; such instances were probably common.

In the course of a single season, one male was observed mating with eight different females, one male with seven, one male with six, two males with three, one male with two, and five males with only a single female each. In contrast, three females were observed to mate with three different males during the season, three females with two males, and three major females with only one male. These numbers of matings recorded obviously depended to a large extent on how long the individuals were observed, and this varied greatly between individuals.

Plate 8.2   A male ostrich scrutinizes his eggs, of which he has different probabilities of paternity.

3. As shown in Chapters 6 and 7, although minor females' eggs were somewhat more likely than major females' eggs to survive until incubation (62% *vs* 54%; Table 7.4), they were considerably less likely to be incubated (61% *vs* 98%; Tables 6.3 and 6.5). Therefore, the reproductive value to a male of a major female's egg is greater than that of a minor female's egg.

Table 8.4 shows the probable reproductive value to a male of mating with a major and a minor female, by combining the data outlined in (2) and (3) above, and assuming that there was no difference in probability between major and minor females of laying an egg following a mating. It can be seen that there was a four-fold difference in the value of matings depending on the category of females.

Since matings are not of equal value, but are of quantifiable value, the curve of mating rate against nesting stage (Fig. 8.5) needs to be modified. For the matings which occurred during the observation periods contributing to this graph, it was known which were with major and which with minor females. The former have each been given a value of 4 and the latter a value of 1. The mating efficacy is the sum of these scores, and is plotted in Fig. 8.9: the mean mating efficacy of a male during the first 10 days after the start of his nest was far higher than before or after this period. The

Table 8.4

The reproductive value to a male of matings with major *vs* minor females

	Mating with	
	Major female (%)	Minor female (%)
1. Probability of paternity of any egg laid	92	32
2. Probability of egg's surviving until incubation	54	62
3. Probability that egg will be in central incubated group	98	61
Overall relative 'value', i.e. 1 × 2 × 3	49	12

peak is even more marked than in Fig. 8.5. Although some males who had lost nests mated fairly often (as shown in Fig 8.5), they were most unlikely to mate with a current major female (as shown in Table 8.3), and therefore their mating efficacy was low.

If matings with major hens were more valuable to males, one would expect that they would try harder to mate with major rather than minor

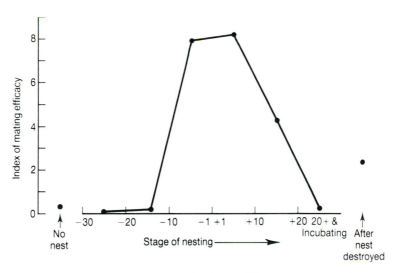

Figure 8.9   The relative mating efficacy of males at different stages of nesting.

Table 8.5

The mating success and effort of males with major females and other females

	Partner		
	Known major females	Other females	
Number of occasions on which the kantle display was followed by copulation	18	27	N.S.
Number of occasions on which the kantle display was not followed by copulation	9	12	N.S.
Number of occasions on which copulation was not preceded by a kantle display	5	14	
Mean number of display waves if a kantle display was performed	38.4	33.7	N.S.

'Other females' in this table includes both known current minor females and females known not to be current major females but not known to be currently laying as minor females.

females. In fact, Table 8.2 shows that the mating rate for each was similar. Table 8.5 shows that the probability of the success of the males' courtship display was also similar for major and minor females. So also was the average number of waves they used in their kantling courtship display to each, despite wide variation. Thus there was no clear indication that males did in fact value copulations with major females more highly.

8.4 Areas of Conflict with Major Female

There were two areas of potential conflict between the male and the major female. The first concerned the amount of time spent guarding the nest before incubation commenced and which of the pair acted as guard. The second concerned the time at which incubation should commence.

a) Attending the nest. If a nest is left unattended, it runs an increased risk of destruction by predators and the eggs could overheat (see Sections 5.3 and 5.4), both of which reduce the reproductive success of both male and major female. If the female spends time guarding the nest, she is deprived of feeding time, which may well be necessary for her to continue to produce eggs; the number of extra eggs she 'loses' as a result probably does not depend on the stage of nest growth. On the other hand, if the male spends time at the nest, his lost feeding time probably has no immediate effect on his reproductive success. However, by being tied to the nest, he also loses mating opportunities with minor females, particularly those two-thirds of the minor females in Table 8.3 who are not laying in his nest; the extent of this 'loss' again probably does not depend on the stage of the nest. The major female does a little better if the male attends the nest rather than her doing so, because she can then possibly lay more eggs, whereas her mate's reduced number of copulations with minor females does not affect her own reproductive output. The male does much better if the major female attends the nest rather than him doing so for two reasons: (1) he can use the free time for mating promiscuously with other females and (2) because the female is tied to the nest, she is less able to mate with other males (at the time of day when those other males are most likely to be away from their own nests and looking for females). Thus he increases his monopoly of paternity of the highest-value eggs his major female lays, even if she lays one or two fewer eggs. Achieving a monopoly at an early date is important because the earlier this is achieved the more eggs will be fathered.

Because minor hens are also laying in his nest, the male's genetic stake in the nest increases at a faster rate than that of the major female and on average it is greater than hers from day 4 onwards (Fig. 8.10). Therefore, in theory, the male will be at a disadvantage when in disagreement with the female over who should attend the nest, because he has more to lose if neither of them attends the nest. The difference in their subsequent probable reproductive output if the nest is destroyed is unknown. One would expect that, other things being equal, the male's attentiveness to the nest would generally be greater than the female's, and particularly so between days 16 and 20, when the differences between the genetic stakes of the two birds are greatest, both proportionately and absolutely. Figure 4.5 bears out this expectation, especially on the major female's non-laying days. It can be seen from Fig. 8.10 that it is on the major female's non-laying days that the differences between the two birds' stakes are greatest.

The picture is complicated by the following facts. First, unlike the male, the major female is compelled to return to the nest at 2-day intervals at minimum in order to lay another egg. Second, the territories are so large that an ostrich cannot necessarily see whether its mate is attending the nest or not, which reduces the scope for bluff. Third, guarding against diurnal nest predators is safer and more effective than guarding against nocturnal nest predators.

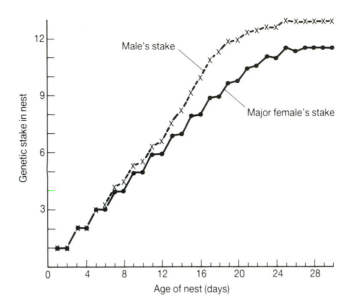

Figure 8.10   Changes in the genetic stake in the nest of the male and the major female with time. Egg numbers are taken from Fig. 4.7. The male is assumed (from Table 8.3) to have a 92% probability of paternity of his major female's eggs, and a 32% probabiity of paternity of minor females' eggs. When there are more than 19 eggs in the nest, it is assumed that enough minor hens' eggs will be expelled to bring the central viable group down to 19; the doomed surplus are deemed not to contribute to the male's genetic stake.

b) *Commencing incubation.* A second area of conflict that arises between the male and the major female is when incubation should begin. For as long as the nest is in existence, there is a risk that it will be destroyed by predators and the whole clutch lost. The more eggs the major hen lays, and therefore the more she delays the onset of incubation, the longer the nest is in existence and therefore the greater is the risk that it will be destroyed. On average, there is an optimum number of eggs that the major hen should lay – if she lays fewer and yet her nest survives, she hatches fewer young; if she lays more (and so takes longer over it), her chances of hatching any of them decline. Thus the product of the number of eggs and the nest's survival probability would be expected to reach a peak and then fall. Figure 8.11 shows this curve for an average major hen, laying as indicated in Fig. 8.10, and subject to a daily nest predation risk of 0.06 (Section 5.3).

However, the curve for the male would not be expected to coincide with that of his major female, because of the presence in the nest of increasing numbers of minor hens' eggs. Also plotted in Fig. 8.11 is the curve against

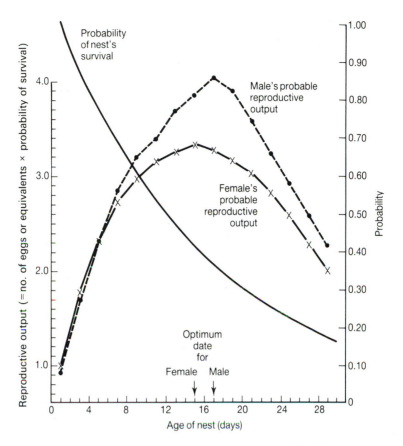

Figure 8.11   The optimum time for a male and his major female to commence incubation in an average nest. The curves show, for each day during the laying period, the probable reproductive output from the nest, assuming a daily nest predation risk of 0.06, and using the data in Fig. 8.10 on the increase in the birds' genetic stake in the nest.

time of the probable reproductive output of the male from his nest; it is based on the same laying rates and risks of predation as those for the major female, and the same assumptions about paternity as in Fig. 8.10.

As shown in Fig. 8.11, for the two birds caring for an average Tsavo nest subject to the daily risk of predation calculated earlier, their probable reproductive outputs from the nest would be expected to peak at about the same time, apparently on days 15–17. This would not necessarily always be the case because the timing of the peaks depends strongly on the variable and largely unpredictable risks of predation on the nest and, for the male, on the rate of accumulation of minor hens' eggs in the nest.

If, at a particular date, male and major female might benefit differentially by either starting or postponing incubation, this might be expected to show up as a difference in the temperature at which each was keeping the eggs in the nest. Readings from fibreglass temperature recording dummy eggs were obtained and analysed to determine whether this was indeed the case (the mode of operation of these eggs is described in Section 2.4). Figure 8.12 shows the temperatures attained by such an egg in an ostrich nest around the start of the incubation period. The periods at which the male and major female were at the nest are indicated, as revealed by time-lapse photography.

There is no clear sign in Figure 8.12 that one bird was in general keeping the eggs in the nest warmer than the other, and the following would appear to account for what was happening. The female was driven off the nest in the middle of the first day so that the already-warmed egg could be put in it. She returned soon after and remained there for the rest of the daylight hours. Her presence probably contributed to the constancy of the temperature of the eggs during that afternoon. Although our time-lapse photography could not be used after dusk, the temperature trace suggested that she departed around that time. Certainly, no bird was at the nest at daybreak, although the female put in a brief appearance a bit later. She was unusually absent on that day (10 September), and the male undertook a greater share of nest attendance. The temperature of the eggs rose as did the air temperature, both while the male was present at the nest and during his subsequent absence.

The nest was apparently unattended throughout the night, and the temperature of the eggs dropped steadily. The arrival of the female at the nest on the morning of 11 September coincided with the beginning of a steep rise in egg temperature; however, it is notable that the temperature of the eggs fell steadily during the afternoon from its peak at 13:00 h, despite the fact that the female was sitting on the nest. Thus she had not, at this stage, begun to incubate the eggs. We cannot be sure at what stage during the night she left the nest, but at dawn on 12 September the nest was unattended and the eggs almost cold. Again the hen's presence during the middle of the day on 12 September was associated with an initial rise and then the start of a decline in the temperature of the eggs. Thereafter, the presence of the male at the nest during the late afternoon kept the eggs at a high temperature until dusk. The temperature trace implies that he was present throughout the night, although it appears that for perhaps 3 h the eggs were uncovered. None the less, at dawn on 13 September, for the first time, the egg began the day at a temperature almost high enough for incubation to commence. It was striking that when the egg was removed at noon on 13 September, all of the eggs in the nest were distinctly cold

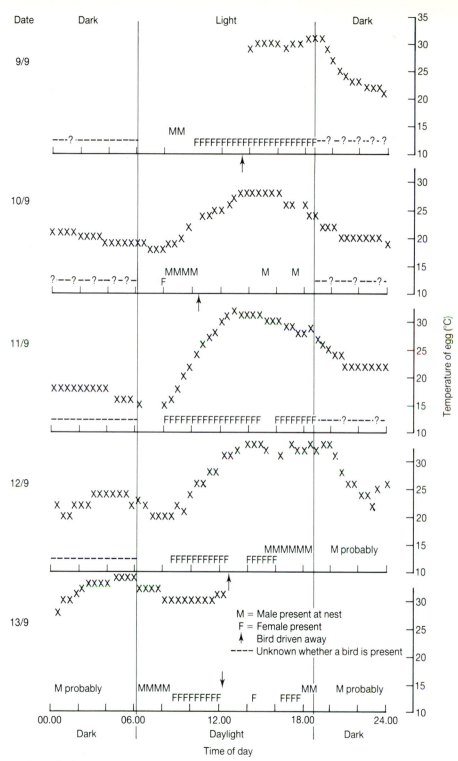

Figure 8.12 The temperatures over 4 days of a temperature-recording egg in a nest around the start of incubation. The presence or absence of the male and female birds is also indicated.

on their undersides, although their tops were warm; subsequent examination showed that embryonic development had not begun.

Egg temperature recordings obtained in a similar way at a second nest round the start of incubation showed temperatures being maintained at between 32 and 37°C for 24 h, as a result of the near-continuous presence of the female at the nest around the middle of the day, and of the male early and late during the day and at night. It was again noteworthy that at this stage the bottoms of the eggs felt completely cold. Obtaining further temperature recordings from this nest was thwarted by the fact that the temperature-recording egg was pushed out into the outer ring, where exposed to the elements it underwent marked daily temperature fluctuations.

It would appear that incubation can only get underway when the male bird is present at the nest throughout the night, thus maintaining a high egg temperature. There is little difference in the middle of the day between the temperatures of eggs in pre-incubating and incubating nests, but both are lower than the temperatures of the eggs left in unattended nests (Fig. 5.1).

The resolution of possible conflict between the sexes over when incubation should begin is discussed in Section 9.4.

9

Discussion:
The Evolution and Maintenance of the Communal Nesting System

Ostrich   Cassowary   Emu   Rhea

The breeding system of the ostrich has been shown to be both varied and complex. It is highly unusual among birds. It is therefore worth asking how does it all come about? This Chapter will examine what selective pressures are operating within the ostrich communal nesting system to influence the choices that individual birds make at different stages of their nesting. It is the consequences of those choices that natural selection acts upon: selection will enhance the frequency in future generations of those genes which make individual birds more likely to make choices that on average have beneficial reproductive consequences.

It is particularly clear that in a complex breeding system such as that of the ostrich, any one bird's choices are affected by other birds' choices (for example, selection of a mate is a two-way process). So, too, are the reproductive consequences of those choices. The ostrich breeding system consists of many individuals interacting, in some cases benefiting to a greater or lesser extent from each others' actions, in some cases being disadvantaged by them. Cooperation, competition, manipulation and compromise are all features of the system.

Close examination of the system and of the choices enables us to understand how and why it is maintained, through the effects of diverse selective pressures. But what factors have predisposed or caused the ostrich communal nesting system to develop in the first place? A number of factors will be examined – some intra-specific (size, sex ratio), some inter-specific (predation) and some environmental (food, climate) – to try to see how they interact.

It is worth looking also at the other large relatives of the ostrich to see to what extent conclusions and generalizations from the ostrich study are relevant to them, and to seek reasons for the differences observed. Finally, the ostrich communal nesting system is put into context as a complex example of what is now known to be the widespread phenomenon of cooperative breeding in birds.

9.1   Choice of Mating Partner

The data on courtship displays indicate that both sexes exercise a consider-able degree of choice in whether or not to copulate with a particular individual of the opposite sex. It was shown in Section 4.1 that only about 68% of courtship displays were successful, and that the major cause of failure was the female's refusal to cooperate. None the less, the same female might well be seen mating with a different male within a relatively short time period. There are probably a number of factors that influence her decision whether or not to mate with a courting male, such as whether or not she is his major hen, whether he has a nest, whether she has mated with him already, and his quality.

One theoretically possible way of assessing the male's quality is by judging the performance of his display. The kantle display is clearly an energy-consuming display, and it may continue for as long as 3 min; therefore, how long it continues might provide the female with a rough indication of the stamina of the male performing it. Poor-quality males presumably do not have the stamina to maintain the display for as long as high-quality males and the duration of the display cannot be faked by the performer. In fact, Fig. 4.2 shows only a slight non-significant indication that females overall were more likely to mate after a long, rather than after a short, display. Many of the kantle displays observed were given to the male's own major female. It would be too late for it to be useful to a major female to assess the stamina of her mate once they have an established nest, since the two have already selected one another as longer-term partners, and this point may confound the data in Fig. 4.2.

Factors other than the length of the display must determine the female's response, but it is not clear what these factors are. There were no signifi-cant differences between the success rates of different individual males (although this may be partly because the samples were small), but a male who had, or was about to have, a nest was more likely to be successful in his courtship displays than males without nests.

No evidence was found (Section 8.3) of males valuing copulations with major females above those with minor females, despite the fact that the latter copulations were only about one-quarter as likely to result in his own progeny (Table 8.4). A male can obviously recognize his own major female, and almost certainly recognize a considerable number of different minor females. Thus the apparent lack of preference between them is not due to his inability to distinguish between them, but is probably due to two possible confounding factors. First, he has a virtual monopoly over copulations with his own major hen; therefore, an extra copulation with her is unlikely to produce a substantial, if any, increase in his reproductive output through her, provided that almost all the eggs she lays are fertile. An extra mating with a minor hen, although it does not increase the probability that she will lay more fertile eggs, could appreciably increase the chances that one or more of them will be fathered by him, depending on the frequency of her matings with different males. His most valuable matings would be those with another male's major hen, but there were insufficient observations of these to indicate whether they are also the most sought after.

The second confounding factor masking any preference for matings with major hens would be a male's ability to assess his own probability of success. It may be that whether or not he performs his courtship displays is influenced by knowledge of whether or not they are likely to be successful with any particular female. She can control whether or not they mate, and she may also indicate to the male that it is not worth while for him to display to her; it is unclear whether or how she indicates this.

In selecting a longer-term nesting partner, males and females may seek rather different characteristics. We would expect the female to select a male who will provide her offspring with a high quality of genetic endowment, and who will provide conditions favourable for successful hatching. Selection on the basis of the size and quality of the male's territory is advantageous for two reasons. With the high degree of territorial competition among males, a male with a larger territory is likely to be a higher-quality male. The environment within the territory may be more or less favourable for nesting, depending on the extent and distribution of cover and of predators, but in general a larger territory provides a greater range of choice of nesting sites. In addition, the female ostrich has to select a male who will be attentive to the nest and young, and once she has laid her first egg, she will be able to tell how attentive he is likely to be. At this stage, if she feels that the male she has paired with is inadequate, she can find herself another partner. A critical point here, and one on which there is little information, is how wide is the choice of possible partners.

When selecting a female to be his major hen, a male would be expected to base his choice on quality, how well she will care for the nest and how many eggs she will lay. Again, we do not know how wide is the choice. Nor is it clear whether males are able to measure any of these three characteristics in advance of nesting, though they probably can to some extent. If a male is intending to select a hen he has attempted to nest with previously – whether in the same or an earlier season – he may have information regarding her attentiveness at a nest. On the other hand, females may be able to deceive males as regards the number of eggs they are still able to lay. A male achieves his peak reproductive output if his major female lays a relatively large number of eggs in their nest – the optimum number will depend on the rate of predation on nests. A female who is still able to lay only a few eggs may be prepared to become a major hen, depending on the energy and predation costs involved. But if she is only still able to lay a few eggs her partner would have done better to have chosen a different hen as his major female. There are no data available on how great is the choice available among those birds prepared to be a major female, nor is there any obvious way by which a male can determine how many more eggs a female ostrich will lay.

9.2   When Should Males Start a Nest?

As shown in Section 5.6, the starting of new nests in the study area was spread over a period of about 3 months. By the time the first nests were begun, all males in the area had established their territories and were displaying to the females. All of the males appeared to be physiologically ready to nest, yet some actually started considerably earlier than others. A male has – theoretically at least – a choice of starting date. Let us now examine the possible factors that influence when a male might start his nest.

A male can only start his nest when a hen is prepared to be his major female, and he will do better if minor females are also prepared to lay eggs in his nest. Thus the spread of dates when it is *possible* for a male to start a nest depends on the laying season of the females; and within that season, the *optimal* time for him to start depends on the degree of synchrony of egg laying by those females and on the times that other males begin their nests. Nesting success and chick survival may also be affected by the date the male chooses to begin nesting.

Figure 8.9 shows that at the time a male begins to nest, his mating efficacy is far higher than during the rest of the breeding season. We noted at least a 20-fold increase in his mating efficacy, many times that of non-nesting males, and he was at his peak for about 30 days of the 10-week period during which he mated. If his nest survived, his mating efficacy fell to zero when incubation began. One would expect a male to be under some selective pressure to start his nest at the end of the season, after first enjoying a couple of months of moderate mating success. But not all males can be last, and a late male can only benefit *because* there are some early nests containing eggs which he has fathered.

Presumably, delaying starting a nest increases the risks of not finding a major hen, although it is not clear whether potential major females are ever in short supply. In addition, starting a nest late gives a male no opportunity of a second nesting attempt if his nest is destroyed. Under the conditions in Tsavo where most nests were indeed destroyed by predators, and provided that second attempts are possible, the optimal strategy for a male is probably to start a nest relatively early, provided there are females available to produce eggs which he has fathered.

On the other hand, the very first male to start nesting may have a lower reproductive output than males starting nests slightly later, because the latter may have fertilized some of the first eggs of the season laid in his nest; the increase in a male's mating efficacy (Fig. 8.9) begins some 2–3 weeks before there are any eggs in his own nest, and it is probable that during that time he is fathering eggs which are being laid in other males' nests. Clearly, more information is needed on the mating rate, the egg-laying rate and their fertilization in relation to the time that his nest commences. However, it is also clear that the number of rival males, the availability of laying females and the rate of nest predation all influence the optimal date on which a male should start nesting.

It is possible that a male can be ensnared into starting a nest sooner than is in fact optimal for him to do, if his major female starts to lay in a scrape in his territory. The male's optimal date for nesting may well not coincide with that of the major female, because of the promiscuity of matings and the egg dumping by females. However, once his major female has started their nest, even if before his optimal time, his best strategy is to assist her; otherwise he will lose the eggs that will contribute most to his reproductive output. Whereas he will probably determine when the incubation period starts (Section 9.4), his major female will probably decide when to start nesting, as she controls egg production. Physical or physiological factors decide which bird can win which conflict of interests.

9.3 Choice of Nest in which a Female Should Lay

As shown in Section 7.1, laying in more than one nest can enable a female ostrich to lay more eggs than she otherwise would. However, in addition, since there is a considerable risk that any nest will be destroyed, there will be some selective pressure in favour of distributing eggs in several nests (i.e. avoiding putting all one's eggs into one basket). This selective pressure will be the stronger the fewer reproductive opportunities there are available to members of a species, and therefore it will be relatively weak in the case of birds as long-lived as ostriches, where a female may well reproduce for 30 years or more. In general, if a female produces, say, 15 eggs per year for 30 years, her average lifetime reproductive success will be the same whether she lays them in only 30 nests or in 100 nests; but her chances of leaving no offspring decrease the more nests she lays in. It would be interesting to construct a theoretical model relating the strength of the selective pressure against putting all one's eggs in one basket to the longevity of the individual and the risk of nest predation.

If a female lays in more than one nest, which nests should she choose? As shown in Section 7.2, minor females' eggs enjoyed better prospects of survival before incubation than major females' eggs, because they were generally laid in nests that were older and which had therefore to survive a shorter period during which they might be destroyed by predators. It is not clear how much this later laying was due to selection by the minor hens of older nests, and how much to poor knowledge of where the more recent nests were. Presumably, it takes some time before a female finds or is shown a male's nest, but time-lapse photography showed that several females had usually visited the nest at some time, before any minor hens had started to lay in it, and so actual selection appears probable.

Another factor which a minor hen might take into account when deciding which nest to lay in is the risk that her eggs will be pushed out into the doomed outer ring (Section 6.2). This depends in part on the total number of minor hens' eggs laid in that nest. If the nest is nearing incubation, and if it contains only the major hen's eggs, then eggs are less likely to be expelled than if the nest is still relatively new but already contains the eggs of several minor females. Presumably, a minor female is able to tell whether only a few or many other minor females are laying in a particular nest. By using this information, combined with an inspection of the eggs in the nest and observation of the male's or the major hen's behaviour, she may be able to determine when incubation is likely to commence. How much use she can make of this information depends not only on her mental abilities, but also on how wide is the choice of different nests available for her to lay in – at times they may well be in short supply.

It can be shown that in most cases, a minor female is better off if other minor females do not lay eggs in the same nest that she has chosen, and that the second minor hen is the one who makes the greatest difference to the chances of her own eggs being incubated. Whereas the major hen benefits from having one or more minor hens laying in her nest (Sections 6.1 and 6.2), but scarcely benefits further from the laying of more minor females, a minor hen suffers from the laying of subsequent minor females, particularly the first of these. She can probably do little to deter them, but she should avoid helping them to find the nest in which she is laying. We might therefore expect to find more aggression between minor females than between a major female and a minor female.

If there is a choice of nests in which to lay, it is possible that a minor hen would do better to lay in the one in which the major hen's eggs are most similar to her own, thereby reducing the chances of her eggs being pushed out into the outer ring. There was no indication that nests in which the eggs of the major hen were, for example, large (or elongated), generally contained large (or elongated) eggs laid by minor hens. However, the discrimination among eggs by the major hen is probably based on features other than, or perhaps as well as, size and shape, and there are no data on other aspects of similarity. Very abnormal eggs, such as exceptionally small ones, those without a surface sheen and fibreglass dummy eggs, were preferentially expelled, suggesting that the choice of eggs to be rejected by the major female produces selective pressure towards a general conformity among ostriches as regards egg characteristics. However, environmental (Section 5.4) and physical and physiological factors (such as breakage and respiration considerations) may impose such conformity, even in the absence of active selection by major hens. One might expect that there would be selection pressure on the major hen to lay eggs easily distinguishable from those of the minor hens; however, this is not the case for two probable reasons. First, major hens are often minor hens at other nests, where their eggs probably need to be indistinguishable from those of the major hen at that nest. Secondly, major hens are able to discriminate sufficiently well (Section 6.2) between their own eggs and those of other ostriches.

One aspect of the choice of which nest to lay one's eggs in is when to lay in one's own nest, i.e. when should a female who has been laying as a minor female in one or more other females' nests start to lay in her own nest as a major female. Figure 9.1 shows in diagrammatic form the probable total reproductive pay-off to a female who switches from minor to major strategies at each stage in her laying series. The assumptions being made are: (1) that her laying is suppressed when she has 11 eggs of her own in her own nest and starts to incubate them (Section 6.3); (2) that a major hen's egg is 1.4 times more likely to survive than a minor hen's egg (Sections 6.2 and 7.2); and (3) that there is a two-egg cost of being a major

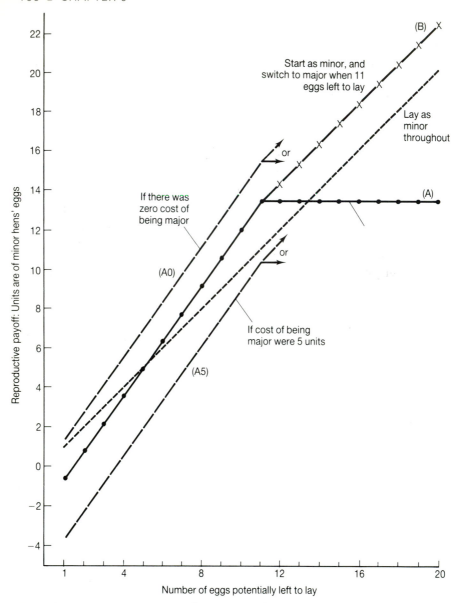

Figure 9.1   The reproductive pay-off for a female of adopting a major or minor strategy at different stages of her laying series. Units are minor hens' eggs; major hens' eggs each have a value of 1.4 minor hens' eggs. Line (A) ●——●——● assumes that a major hen's egg laying is suppressed after she has laid 11 eggs as a major hen, and that there is a 2-egg cost of caring for a nest. Lines (A0) and (A5) show pay-offs if there were a cost of 0 and 5 units (minor hens' eggs) respectively of caring for a nest. The optimal strategy (B) is to lay as a minor female until 11 eggs remain to be laid.

hen, because of reduced feeding (Section 7.2) and perhaps an increased risk of predation. The costs of being a major hen cannot yet be quantified; however, they are assumed to be small and independent of the number of eggs laid.

It can be seen in Fig. 9.1 that it is worthwhile for a female to switch from being a minor to being a major hen when she has 13 or fewer eggs left to lay, even though this results in her laying fewer eggs in total than she otherwise would. It would be even more worthwhile if she waited until she had only 11 eggs left to lay. However, she cannot guarantee finding a male ready to start a nest at any particular instant, and presumably the chances of her subsequently finding such a mate will influence her decision when to start her own nest. Even when she has only 6 or 7 eggs left to lay, it is worthwhile her becoming a major female; however, her prospective mate will suffer a reduction in his reproductive output if he starts a nest with her rather than with a female who has a full number of eggs left to lay. It is not clear whether or how he can detect how many more eggs she will lay, but it was apparent that not one major hen laid fewer than eight eggs in her own nest. It is unlikely that the cost of being a major female is as great as five minor hen's eggs, but if it were, all females would apparently do better to remain minor females throughout their laying period (Fig. 9.1). If there was no cost to being a major female, it would always be worth trying to become one.

The above calculations on the reproductive benefits of being a major hen are based on the population characteristics of ostriches in Tsavo. It is likely that elsewhere the relative benefits are very different. For example, the costs of being a major female probably vary greatly, since they depend on the abundance of food and the risk of predation. Equally, the greater the proportion of minor hens in the population, the greater are the relative benefits of becoming a major female, and vice versa; therefore, the system tends to stabilize. The strategies of being major or minor are not strategies with equal reproductive pay-offs. Rather, the evolutionarily stable strategy (Dawkins, 1980; Maynard Smith, 1976) is probably a mixed one – being minor at first, becoming major at a moment which is influenced by a variety of factors, and being able to revert to being minor in the event of a late disaster to the nest.

9.4   Cessation of Laying and Start of Incubation

Ostriches are indeterminate layers, i.e. the number of eggs a female will lay is not determined at the start of the laying period, but depends on factors during the accumulation of eggs. Just what all these factors are is not known, but they almost certainly include stimuli related to the number of eggs in the nest. Information from ostrich farms suggests that an isolated pair of birds will usually start to incubate when the female has laid about 16 eggs; that a trio (one male and two laying females) will usually lay until there are about 16 eggs in the nest and then start to incubate; that a group consisting of a male and three or four laying females will usually overshoot and start to incubate when there are a couple of dozen eggs in the nest; and that a female whose eggs are removed daily will generally lay about 30 eggs before ceasing laying (B. Strydom, pers. comm.). The maximum number of eggs laid in captivity in a year is over 100 (Smit, 1963). In the wild, indications of the feedback effect of egg numbers come from the non-significant negative correlation (Section 4.4) between the numbers of major hens' eggs and minor hens' eggs in nests surviving until incubation. Contributing to this correlation in 1979 was Nest 6, in which no minor hens laid any eggs until day 30, and perhaps in consequence the major female laid 17 herself, more than any other major female did in her own nest. It is not known how the feedback works in ostriches, nor how rapidly it can operate in response to rapid changes in the number of eggs in a nest. The maintenance in evolutionary terms of a system whereby it appears that minor females may be able to influence the major female's reproductive output and behaviour, needs investigating.

It was suggested in Section 8.4 that the risk of nest predation was a determinant of when incubation commenced. The problems of measuring daily nest predation risks were described in Section 5.3. The calculations assumed a constant daily risk, such as would apply if the distribution of nest predators was random over the environment and constant over time. But almost certainly it is not random – for example, hyaena dens and Egyptian vulture nests cause these nest predators' activities to be more concentrated in particular places, such as along particular routes, than in others. As a result, the daily risk of destruction of a 2-week-old nest is probably less than that of a newly-started nest. There were insufficient data to determine this likely decline. This does not alter the principle of there being an optimum number of eggs, but it does affect just what that optimum number will be.

The optimum time to start incubating is not necessarily the same for the male as for the major female, because the minor hens' eggs which are laid in the nest are not laid at the same constant rate as the major hen's eggs. The major female would be expected to prefer to commence incubation earlier than the male if she had few eggs left to lay, or if the male did not monopolize matings with her, or if the minor hens' eggs were being laid in her nest at an increasingly faster rate as the laying period progressed.

On the other hand, the male would be expected to prefer an earlier start than his major female if a large number of minor hens laid in his nest early on, and if the risk of nest predation was relatively high. We found that, in many nests, no minor hens' eggs appeared for several days after the start of the nest (Section 4.3), perhaps because of the difficulty of finding it, and perhaps because an early-stage nest is a less safe place than older nests to entrust eggs to. But in some nests, particularly those which were immediate re-starts after the destruction of a previous nest, several minor hens' eggs appeared within the first few days. The divergence between the optimum times for males and females is kept narrow by the limited number of eggs which can be incubated, and by the major female's ability to discriminate between eggs (Section 6.2 and Bertram, 1979) and to expel as surplus some of those which she herself had not laid.

There was no clear evidence of conflict between a male and his major female over when incubation should start. This could be because (1) the individual birds' reproductive interests coincided exactly; (2) the birds' behaviour has not evolved to be sufficiently finely tuned to take into account such changes in fitness through alterations of strategy; (3) the measures of attentiveness (temperature, times at the nest) were too crude to detect conflict; (4) that any conflict would be so one-sided that it does not take place. Explanation (1) was clearly not always the case, (2) is an 'explanation' of last resort, (3) is very probable but unrewarding, and (4) will be examined in more detail below.

The temperature of unattended eggs fluctuates noticeably over the 24-h cycle. During the day, heated by the sun and the air, egg temperatures rise considerably, up to and above that necessary for incubation. During the evening, they fall steeply, and for most of the night they remain low. The female's presence on the nest during the day keeps the eggs cooler than they would otherwise be.

Figure 9.2 shows the approximate temperature profile of an ostrich egg in a hypothetical nest where the male had embarked on incubation but the major female was reluctant to do so. During the night, the male has warmed the egg to the presumed incubation temperature of around 36°C. At 09:00 h his major female comes to relieve him for the daylight period. If she merely shaded the eggs rather than keeping skin contact with them, the egg temperature would drop at a rate dependent on the temperature difference between the egg and the air; ostrich egg cooling rates have been calculated to be approximately 0.4°C per °C temperature difference between egg and air per hour (Drent, 1975). Because ostrich eggs are large, they cool relatively slowly. Within $2\frac{1}{2}$ h, the daytime air temperature is above that of the egg, which consequently starts to warm up again. The major female cannot prevent this, and in her absence the egg would warm

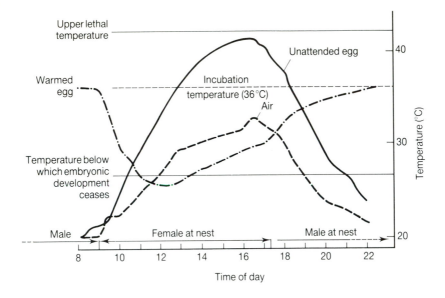

Figure 9.2   The temperatures in a hypothetical ostrich nest, based on data and calcu-lations. ---- indicates the shade air temperature measured on a typical day during the breeding season in Tsavo. —·—·— indicates the temperature course which would be followed by an egg if (a) the male departs from the nest when the eggs are at the temperature necessary for incubation at about 09:00 h; (b) the female tries to keep it as cool as possible by shading it; (c) the male returns at 17:15 h and warms it by sitting closely on it. ———— indicates the temperatures reached by an unattended egg exposed to full sunlight, even if starting the day cold.

up even faster and could possibly overheat. The male returns at 17:15 h, quickly warms the eggs the few degrees necessary to reach full incubation temperature, and then keeps them at that temperature throughout his 16-h nightshift. It can be seen in Fig. 9.2 that for only about $2\frac{1}{2}$ h of the day is the egg temperature below that at which embryonic development is held in abeyance. Note that the temperature at which embryonic development is said to slow down or stop is that established for domestic chickens' eggs (Lundy, 1969); the relevant temperature for ostrich eggs is unknown, as is the length of time that an ostrich egg must be kept constantly at the incubation temperature before embryonic development can begin.

Thus if the male starts to incubate the nest, there would appear to be very little that the female can do in practice to prevent it, because she attends the nest at those times of the day when significant cooling is impossible, and because the very large eggs cannot be cooled quickly. The most she can do is temporarily slow down the embryonic development, but there is little benefit to her in doing so once such development is under way. She could spend much longer at the nest, instead of the male, keeping the eggs cool; but then she would lose feeding time and be unable to lay many more eggs.

Thus the male is in the best position should a conflict arise between the members of a pair over when incubation should begin. Because the male tends the nest overnight, i.e. the only time that the eggs can be cooled, he can irrevocably initiate embryonic development or he can probably prevent it from starting.

9.5   Incubation and Hatching

The laying of eggs in a nest generally stops within a very few days of the start of incubation, although in a few nests, and particularly in smaller ones, the odd egg may be laid by a minor hen even quite a long way through incubation. The minor hens may continue to lay in other nests without interruption, so how do they know when to stop laying in the previous nest? There are several possible signs they may use. The first is the presence of an outer ring of eggs. As described in Section 6.2, the outer ring is formed by the major hen around the time when incubation begins. The appearance of an outer ring thus indicates unmistakably the advent of incubation. The second sign is the behaviour of the major female. An incubating major hen is reluctant to leave her nest – she does not stand up and move aside for a minor female wanting to lay, as she would during the laying period. The third possible sign is the temperature of the eggs. It is not clear whether this is a cue which the birds are actually able to use, because pre-incubation eggs are often only slightly cooler than those being incubated and because ostriches may not have a suitable temperature-measuring mechanism.

It is clearly in the interest of any female not to lay in a nest in which the eggs are already being incubated because the egg would not hatch until after its companions. This would not necessarily be the case if there were an accelerating or synchronizing mechanism influencing embryonic development and hatching, as has been shown to exist in quail and other galliform birds (Vince, 1969) and in rheas (Bruning, 1974). There is as yet no evidence of such a mechanism in ostriches. Hatching in most wild nests is spread over a period of 3–4 days (Hurxthal, 1979). The incubation period does appear to vary appreciably, probably according to the incubation conditions (Hoyt et al., 1978; M. Goussaard, pers. comm.;

A. P. Koen, pers. comm.), but there is no evidence of communication between eggs influencing embryo growth rate (R. Faust, pers. comm.; Bertram, unpublished observations). There may be some communication which helps to synchronize the precise time of hatching.

The ability of a late-laid egg to accelerate its embryonic development would have a considerable influence on the ostrich communal nesting system. It would be beneficial to the major hen who might therefore be able to continue to lay eggs even after incubation had started, and it would be beneficial to the minor hens who would therefore have a wider choice of nests in which to lay. Why has this not evolved, as it has done with some other species? The most likely reason is that ostrich embryonic development has already been maximally shortened during evolution. The ostrich incubation period is very short in relation to the size of the egg (Hurxthal, 1979; Lack, 1968; Rahn and Ar, 1974), and it is likely that this is in response to the high rate of nest predation – the shorter the incubation period, the less is the risk that the nest will be destroyed. A second possible factor is the reduction of kin selection pressure in favour of embryonic acceleration or retardation, which would be found in most nidifugous birds which exhibit the phenomenon. In a quail nest, for example, all the eggs are likely to have been fertilized by the same male and laid by the same female. As well as the individual advantage to each chick to hatch at the same time as its companions, kin selection (Hamilton, 1964) may favour mechanisms by which chicks help their full siblings to hatch at the right time. The chicks hatching in an ostrich nest are on average much less closely related to one another (Table 6.1) because of the promiscuity of their parents and the communal laying of their mothers, so that any selective pressure to assist their companions to hatch synchronously must be weaker.

When incubating their nest, as well as during the laying period, the male is generally at the nest throughout the night and the female for the middle part of the day. The female's brown plumage is much more cryptic than the male's, and it is often stated that the plumage of each sex is related to the time each is at the nest. This is unlikely to be the case. During the daylight hours, visual crypticity is likely to be more important than at night, although the latter period is when most predators are active (Schaller, 1972). However, attendance at the nest during the daylight hours is divided almost equally between conspicuous male and cryptic female (Fig. 4.8). The male's colour is not adapted towards crypticity at night – he would be better concealed if he were brown like the female – but is almost certainly an adaptation towards maximum conspicuousness when not at the nest. The dense black spot of a male ostrich can be seen from several kilometres away, by ostriches as well as by human observers.

Much of the males' territorial advertisement and attraction of females depends on their conspicuousness. Females do not need to be conspicuous. Either sex, when absolutely motionless on a nest, is in fact quite difficult to detect, especially because the nest is usually concealed among vegetation or other obstacles. The evolutionary reasons for the allocation of time at the nest are probably nothing to do with conspicuousness but more to do with the time of laying. Ostriches sit inactive at night, and can probably see little in the dark. Travelling to and from a nest means that eggs must be laid during daylight hours. The need for a cautious approach, an undisturbed period at the nest, and time for a possibly delayed departure, mean that laying must take place reasonably near the middle of the day. To have the female at the nest during the middle part of the day reduces the number of journeys that she has to make to it. If the male does not attend the nest in the middle of the day, he is able to mate with other females who enter his territory, some of whom will be searching for a place to lay, or coming to lay there having been shown it already. With the male attending the nest during the long nighttime period, both sexes benefit.

9.6 Factors in the Evolution of the Communal Nesting System

A variety of factors, at different levels, have influenced the evolution of the communal nesting system of ostriches. In this section, let us consider in turn the effects of their large size, the sex ratio, the predation rate, the food supply and the climate, and how these various factors interact with one another. These complex interactions are summarized in Fig. 9.3.

a) Large size. The ostrich is by far the largest extant bird. It is likely that the primary causes of its very large size are predator pressure and feeding efficiency. As a ground-feeding walking bird, the ostrich's size enables it to move and feed relatively efficiently when compared with smaller birds – locomotion is cheaper for larger animals (Taylor *et al.*, 1970). Its size also renders it invulnerable to the smaller predators such as jackals, and enables it to outrun the larger predators such as lions, as well as providing it with the advantage of height in detecting their approach. And larger animals are likely to be better able to survive through periods of adversity than smaller animals.

Whatever its causes, the ostrich's very large size has evolutionary conse-

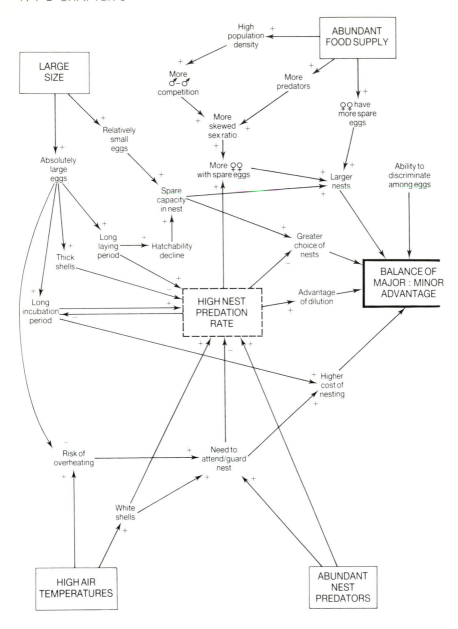

Figure 9.3 The interaction of factors influencing the evolution and maintenance of the ostrich communal nesting system. + and − indicate positive and negative effects.

quences which ramify throughout its breeding system. It obviously precludes flight, as well as ensuring that there can be no aerial predator to threaten adult ostriches, for a sufficiently large predator would also be unable to fly. In general, larger species live longer than smaller ones (Lack, 1968). Ostriches probably live for at least 30 years even in the wild, during which time there is considerable scope for learning by observation or by trial and error about such matters as territory quality, mate selection and nest site selection, all of which influence the communal nesting system.

Large birds lay large eggs, and at a mean 1560–1660 g (Fig. 5.8), the ostrich's egg is the largest birds' egg of all. Large eggs take a long time to be formed, probably for the simple physical reason that additional material is added from their outer surface and the surface area : volume ratio decreases as eggs get larger. In addition, large eggs need very thick shells, for two reasons: first, they have to support large birds moving and sitting on them, and secondly, they are less convex and therefore inherently weaker. Ostrich eggshells are about 2 mm thick. A sample of 31 eggshells averaged 19.8% (range 17.4–21.6%) of the total egg weight; the eggshells of most ground-living birds average a much lower percentage of total egg weight (Lack, 1968). The thickness of the shell presumably leads to a slow rate of production, both in terms of the time taken to lay down the shell and perhaps also of the time needed to mobilize the necessary calcium from the diet or from the skeleton. The result is that ostriches lay at intervals of not less than 2 days, which extends the time needed to complete a clutch.

The large, conspicuous eggs with their thick shells are invulnerable to all avian predators except the Egyptian vulture; thus it is possible – even if inadvisable – for ostriches to leave them unattended for reasonably long periods. Egyptian vultures are rarely abundant, are conspicuous and are unable to plunder a nest particularly quickly, and so ostriches are able to protect their nests even if some distance from them. Because the eggs are so large, they do not heat up or cool down very quickly if exposed to the elements. An ostrich can safely leave its eggs unattended in a hot environment for much longer than could a smaller bird, such as a sandgrouse, whose eggs would be cooked within minutes. Similarly, if an ostrich has to leave its incubating eggs exposed, being large they cool only relatively slowly, and take a long time to chill completely.

In general, large eggs have long incubation periods (Lack, 1968). The duration of the incubation period (6 weeks) influences the optimal clutch size and the cost of nesting, which in turn influences the relative pay-offs of the major and minor female strategies.

Although the largest bird's egg, the ostrich egg is the smallest in relation to the size of the bird that lays it. A mean egg weight of 1600 g (Fig. 5.8)

Plate 9.1   The strength of an intact ostrich egg is demonstrated by the author and his wife.

is about 1.5% of the weight of an average adult female (Smit, 1963; Hurxthal, 1979; A. Milewski, pers. comm.; but note that reliable data on average adult ostrich weights are scarce). This is a reflection of the general allometric relationship across bird species between body size and egg weight, such that smaller birds generally lay relatively larger eggs (Lack, 1968; Rahn *et al.*, 1975; Hurxthal, 1979). At the bottom end of the scale, the egg weight may be as much as 14% of the body weight of smaller bird species.

There are as many evolutionary consequences of laying relatively small eggs as there are of laying large eggs. The most important from the point of view of the evolution of the communal nesting system is that ostriches are able to cover a large number of eggs – about 19 appears usually to be the limit. The ostrich major hen is unlikely to have laid as many as that herself for a variety of reasons: some may be physiological, to do with a bird's ability to gather, store or mobilize sufficient protein, fat or calcium; others are concerned with survival, both because of the increased risk of predation resulting from a prolonged laying period (Section 5.3) and because of a reduction in hatchability of the first-laid eggs with age (Section 6.3). An ostrich can therefore incubate more eggs than she is likely to be able to lay herself, or than it is worth her laying, and thus has spare incubating capacity beneath her. It is this spare capacity in ostrich nests which opens the way for minor females to lay successfully in them, and which sets off the evolution of the complex consequences described in earlier chapters.

Small eggs naturally result in small chicks, which are therefore vulnerable to a greater number and variety of predators than larger chicks. On the other hand, the protective behaviour of the large adult ostriches is effective in at least reducing predation on their chicks.

Ostrich eggs have a relatively short incubation period. The various allometric relationships among bird groups of different aspects of their reproduction have been examined in detail by Hurxthal (1979). While extra caution is needed in extrapolating from normal-sized birds to those species at the upper end of the size spectrum, such relationships show that the ostrich incubation period is appreciably shorter at 42 days than that expected on the basis of egg size (59 days). This unusually short incubation period may be an adaptation to the considerable risk of nest destruction during incubation, and it is probably made possible by the fact that the ostrich chick brain weight is low (Hurxthal, 1979). The brain weight of adult ostriches is also low for the birds' size, compared with other species; it is possible that if the growth rate of neural tissue tends to limit body growth rate of both embryo and hatched chick, then the relatively small egg, short incubation period, rapid growth rate and small brain may all be linked adaptive responses to strong predation pressure on nests and chicks (Hurxthal, 1979).

b) *Sex ratio.* The skewed sex ratio among breeding adults appears to be common among ostriches, as it occurs in the Nairobi National Park (Hurxthal, 1979), the Serengeti (Bertram, unpublished observations) and Namibia (Sauer and Sauer, 1966a) as well as the Tsavo West National Park. The causes of it are not clear. There are no data on the sex ratio of ostriches upon hatching; on ostrich farms it is assumed to be 50–50. There are no clear data on the sex ratio of young birds in the field because of the problems of sexing them (Section 3.1). Subsequently, there are probably three factors that contribute to the relative scarcity of breeding adult males. First, males probably start to breed one or more years later than females; but clearly, in a long-lived species, this alone will make relatively little difference to the breeding sex ratio. Second, as a result of intense male competition, some males may be driven out of the breeding areas; subadult males are certainly chivvied by one territorial male after another but where they go is unknown. Third, the mortality of adult males may be greater than that of adult females, perhaps because of their greater conspicuousness (Section 9.5), perhaps because they are more vulnerable at the times of day they attend the nest (Section 4.5), perhaps because having a higher genetic investment in the nest than the females, the males are more reluctant to leave or more prone to defend the nest at the approach of a predator, and perhaps because in the course of inter-male competition, males are more likely to be injured, exhausted or unwary.

The shortage of adult males means that some females are unable to have mates, and thus are obliged to be minor hens, predisposed to make use of any spare capacity under incubating birds. The extent of the bias in the sex ratio influences the number of minor hens per nest, and therefore the reproductive pay-off to the minor hens, more of whose eggs will be pushed out as the number of eggs in the nest increases.

No cases were noted of a female nesting on her own, probably because it would be almost impossible for her to succeed in hatching any eggs. While the cost of foregoing all feeding for 8 weeks is probably high, the risk of predation to the nest if left unattended during feeding is also high (Section 5.3). If two females were to cooperate in setting up and tending a female-only nest, they would be able to overcome these problems. Yet no such cases were found. It is possible that there is some fundamental biological reason for this. It is also possible that the resultant lack of monopoly of choice of which eggs were to be pushed out would result in conflict and reduce the net benefit of a major hen; on the other hand, if properly organized, each female could on average have 9.5 of her own eggs incubated in the central group and this figure is close to the average major hen's total of 11 (Table 6.5).

c) Predation rate. The rate of predation has a variety of evolutionary influences on the ostrich communal nesting system. As described above, predation may have been one of the factors in the evolution of their large body size, and may contribute directly to the uneven sex ratio. Predation on birds while at their nests influences the cost of nesting, and so influences the reproductive pay-off to major females and males. Predation on the nests themselves may have been a factor in the evolution of the short incubation period and perhaps therefore of the eggs' inability to synchronize embryonic development (Section 9.5). It also results in there being more females who have surplus eggs because their own nests have been destroyed; there are thus more minor females, which in turn reduces the average reproductive pay-off to minor females. Predation on nests produces part of the need to attend the nest, and thus also influences the cost of nesting. The need to guard the nest cuts down on feeding time and therefore probably reduces the number of eggs the major female can lay. It determines whether or not a single individual can care for a nest alone, as with rheas (Bruning, 1974) and most other ratites (Handford and Mares, 1985), or whether two birds are essential. It presumably influences both the optimal nest size (Section 6.3) and any conflict between the sexes over the time when incubation should begin (Section 8.4).

d) *Food supply*. The food supply is likely to have a regulating influence on ostrich population densities (Lack, 1968). The population density influences a variety of aspects of the communal nesting system. One of these is the degree of territorial competition, which probably influences the functional sex ratio, the effects of which are considered above. Territory size probably influences the degree of promiscuity, since females can meet more territorial males in a short time if those males have small territories. The sizes of these determine the maximum distance between nests and so influence the number of nests available for a minor female to lay in; having a wider choice may enable her to place her eggs more advantageously (Section 7.2). At higher population densities, upon hatching chick groups would be expected to meet and merge more quickly, and so probably reduce their numbers lost to predators.

The available food supply is likely to influence the number of eggs that female ostriches can lay. If there is more food, all females will presumably be able to lay more eggs. As a result, because the numbers of eggs per nest or per bird will rise, the proportion of eggs which are minor hens' eggs will increase, and so the relative pay-offs of the major and minor laying strategies will alter in favour of the major hens. More food may also result in larger eggs (Section 5.6) which may affect both the number that can be covered and thus incubated, and the chances of survival of the chicks.

The more food there is available, the less is the time needed for its ingestion, and therefore the more time there is available for guarding nests. The time spent guarding is likely to be one of the factors which, by reducing feeding time, affects the number of eggs that major females can lay in their own nests.

e) *Climate*. The temperature and the amount of sun affect the survival of

unattended eggs in a number of ways. The hotter and brighter the environment, the sooner an unattended egg will overheat. As with the need to
guard against predators, the need to shade the eggs against the sun cuts
down on feeding time and probably reduces the number of eggs the major
female can lay. Temperature is probably the main cause of the decline in
hatchability with age (Fig. 5.2), and therefore affects the number of eggs
it is worthwhile for a major hen to lay in her own nest before starting to
incubate the clutch. The shiny bright white colour of ostrich eggs is
probably an adaptation against overheating (Bertram and Burger, 1981b);
the other large ratites mostly lay green eggs (Bruning, 1991). On an
evolutionary time-scale, high temperatures produce the need for a colour
which renders ostrich eggs more conspicuous and so vulnerable to aerial
predators (Section 5.1), and thus increases both the predation rate on nests
and the need to guard them.

Seasonal changes in climate probably affect the length of the breeding
season, because successful nesting seems to be possible only when there is
no heavy rain. The costs and benefits of the strategies employed are
affected by how synchronized the ostrich population is in its nesting. The
availability of food for small chicks after hatching is probably also
important in determining the ostrich breeding season, and depends on
seasonal climatic changes. Climate clearly affects the type and amount of
vegetation cover; greater grass cover is likely to reduce the rate of
predation on nests.

The productivity of an area, and therefore the vertebrate biomass which
it can support, is clearly affected by climate. In more productive areas, the
ostrich food supply would be expected to be high, but so would the
number of other vertebrate species competing for the same food supply.
Therefore, the relationship between productivity and ostrich population
density is not straightforward. But a high vertebrate biomass results in
more vertebrate predators, in particular (in relation to ostrich nesting)
hyaenas, jackals and lions. It is likely that as the number of predators
increases, the chances of an ostrich nest remaining undetected for the
necessary 9 weeks reduces dramatically. And as outlined already, changes
in the rate of predation on nests have considerable effects on the pay-offs
of the different strategies of nesting ostriches.

9.7 Communal Breeding in Other Ratites

The other ratite species – the larger relatives of the ostrich – go in for
communal nesting in somewhat different forms, usually simpler than that
of the ostrich. The current scanty state of our knowledge of their breeding

systems has been summarized by Handford and Mares (1985). Few generalizations are possible about the group as a whole.

The breeding system of the Greater Rhea (*Rhea americana*) of the pampas of South America has been investigated by Bruning (1973, 1974). The bird weighs up to 45 kg, with males somewhat larger than females. The winter groups fragment as the breeding season begins, and males display and compete with one another. The most dominant is the first to acquire a harem of up to a dozen females who each lay in his nest at 2 - to 3-day intervals for 7–10 days, before moving on as a group to the next most dominant male. A female may lay for 10 different males in a season. The male alone incubates the clutch, starting by the third day. Many nests are abandoned, particularly later in the season, and often after the explosion of surplus uncovered eggs. There is marked synchrony of hatching within a nest, despite some eggs having been laid a week later than others; according to the conditions, the incubation period, which averages 36 days for a clutch, can be shortened to 29 or lengthened to 42 days.

The Emu (*Dromaius novaehollandiae*) from the arid open areas of Australia is another species in which the female takes no part in incubation (Davies, 1963, 1976). The male starts to incubate (for 8 weeks) the clutch of 5–20 greenish eggs before laying is completed. There are indications that more than one female may lay in the nest (Gaukrodger, 1925). Females, at up to 60 kg, are about 10% larger than males.

The three species of Cassowary from northern Australia and New Guinea, are largely solitary frugivorous inhabitants of woodland and forest habitats (Davies, 1976). The female is brighter and considerably larger than the male. Very little is known of their breeding system; they have generally been considered to be monogamous. Crome (1976) reported that one female mated with more than one male each year; each male in succession undertook all the incubation of her clutch of up to 6 eggs.

The three Kiwi species, *Apteryx*, are unlike the other ratites and the tinamous in being nocturnal and monogamous. They lay a clutch of two relatively enormous eggs, up to 25% of adult body weight, which the male alone incubates for 10–12 weeks, attended occasionally by the female (Reid and Williams, 1975).

Few of the more than 40 species of Tinamou have been studied closely. They are chicken-sized, ground-living birds from South America, and unlike the other larger ratites they are able to fly. The females are usually larger than the males. Among Boucard Tinamous (*Crypturellus boucardii*), it appears that a small group of females lays a clutch for one male and then moves on to the next nest; in each case, the male alone incubates the clutch of up to a dozen eggs (Lancaster, 1964). The indications are that laying in a subsequent male's nest may occur in other genera, including *Nothoprocta* (Pearson and Pearson, 1955) and *Nothocercus* (Schafer, 1954).

The breeding systems of the four large ratite groups are summarized in Table 9.1. It must be made clear that good comparative data are very scarce. None the less, a marked feature of all ratite breeding systems seems to be that they are flexible. Thus, even a small sample of nests or studies throws up a wide variety of nest sizes, of adult grouping patterns and of the female's role in nesting.

Among ostrich studies, for example, Sauer and Sauer (1966a) suggested that groups of female ostriches were relatively consistent with a clear dominance hierarchy within them, whereas neither this study nor that by Hurxthal (1979) found any consistency of female groupings. The nests observed by Hurxthal (1979) were in some cases considerably larger than those in Tsavo, as might be expected on the basis of the higher population density of ostriches there.

Generalizations about the species can be unreliable. It is difficult to define the ostrich or rhea system meaningfully in terms of polyandry or polygyny, and caution is similarly needed in drawing comparisons between species. And, sadly, there is an almost complete absence of quantitative data on feeding ecology and behaviour which underlines the birds' social systems.

The ostrich differs markedly from the other large ratites in that females play a large part in caring for the nest. As we have seen, not all females do, but at any one nest the major female does her fair share of guarding the clutch and later incubating it. Although males were found to incubate for 71% of the time, this was because they attended the nest during the inactive night period. Figures 4.3, 4.4 and 4.8 show that the major female and the male took roughly equal shares during the daylight hours, which is the only time when nest attendance precludes feeding or other reproductive activities.

Although unusual among ratites, equal participation by males and females in nesting is of course by no means unusual in birds in general, and nor is participation by the male (but the situation is not necessarily straightforward; see, for example, Van Rhijn, 1984). I have shown in some detail in earlier sections some of the reasons why the ostrich breeding system has evolved and is maintained; it is fruitful to consider why the same system is not found in the other three large ratite groups.

1. Why do most other female ratites not participate in nest care? The answer probably differs with the species in question, but is primarily that the female is not necessary because the male can manage the incubation

Table 9.1

The breeding systems of the large ratite species

Species	Approximate weight of females (kg)	Sexual dimorphism	Clutch size	No. of females laying in nest	Incubation by	Merging of broods	Source
Ostrich	110	Male somewhat larger and conspicuously different plumage	15–36 ($\bar{x} = 25$)	2–6+	Male and major female equally	Yes	This study
Greater Rhea	40	Male slightly larger, similar plumage	20–40 ($\bar{x} = 26$)	10–12	Male only	No, but acquire stragglers	Bruning (1974)
Emu	60	Male slightly smaller, similar plumage	5–20 ($\bar{x} \approx 9$)	1	Male only	No	Davies (1963)
Cassowary	58	Male much smaller and less bright	3–8 ($\bar{x} \approx 5$)	1	Male only	No	Crome (1976)

and the care of the precocial chicks on his own. In the case of rheas certainly, apparently in cassowaries, and just possibly in emus, the female's freedom from nest duties enables her to lay subsequent clutches in the nests of other males. It is probably the high rate of predation on ostrich nests which makes the attention of both birds necessary, and perhaps also the relatively sparse food supply in the arid environment, such that an incubating bird cannot sustain itself with only very brief absences from the nest. There are no quantitative data on the rate of predation of the nests of the other species, but it would appear not to be very high. Whereas two ostriches are essential for their nest to have a reasonable chance of surviving, this seems not to be so with the others.

2. Why do the females of all the other ratite species not lay many eggs in one another's nests? To a large extent, rheas also do. Frequent laying in other individuals' nests requires spare eggs, surplus capacity in nests, knowledge of where those nests are, and acceptance of others' eggs in them. It is likely that among emus and cassowaries these conditions are not met. It may be that food and perhaps mineral resources limit the number of eggs that can be laid, nests may be difficult to find in thick vegetation, and their owners may not be able to benefit as ostriches do from dilution of their own eggs or chicks by others.

3. Why do rheas not go in for egg manipulation as ostriches do? There are two aspects to this. Bruning's (1974) observations appeared to indicate that males were attempting to incubate too many eggs, as abandonment following explosion of a rotten egg was very common. Unincubated ostrich eggs in the outer ring did not rot for months, and I presume that the rheas' eggs which did were eggs in which the embryo had started to develop and subsequently died. Arranging the eggs before incubation began, as ostriches do, might in principle resolve the difficulty for rheas; but this might not be practicable because incubation begins when the rhea clutch is still small, with communication between eggs being required to synchronize embryonic development and hatching.

In the rhea system, the male has no reason to discriminate one female's eggs from another's, since he has mated with all those females. It is not known how his confidence in his paternity would change in relation to early- or late-laid eggs in his clutch.

4. Why do the other ratites not show the marked sexual dimorphism that ostriches do? In a vertebrate species such as the ostrich, in which males are territorial and polygamous, it is far from unexpected to find that males are larger, better armed or more decorative than the females. There are numerous examples in the comparative mammal and bird literature (e.g. Clutton-Brock and Harvey, 1978; Crook, 1965; Emlen and Oring, 1977; Geist, 1977; Jarman, 1974; Lack, 1968; Orians, 1969; Williams, 1966).

Monogamous species tend to exhibit much less sexual dimorphism, as with the emu. Polygamy tends to be associated with reverse sexual dimorphism (Jenni, 1974) and the cassowary system is consistent with such a tendency. It may be that the combination of simultaneous polygyny and sequential polyandry in the rhea in effect annul one another and result in a uniformity between the sexes.

The differences between the ostrich and rhea systems are not very large. In fact, an ostrich female who has consistently been a minor female throughout the breeding season, would in fact be dumping eggs in a number of different males' nests in a way similar to that of a female rhea.

9.8   Communal Breeding in Other Species of Birds

It is not only among the ratites that several females may lay in one nest. There would appear to be two main categories in this regard. The first may be broadly described as brood parasitism, and the second as cooperative breeding; the categories are not necessarily distinct.

Brood parasitism has been well reviewed by Andersson (1983). It is apparently more prevalent in precocial than in altricial species; in the former, the nest is less likely to be occupied by the host female who owns it. The parasite does not contribute to incubation or rearing. The practice of dumping eggs in the nests of other members of the same species has been reported to occur in particular among waterfowl, such as the Carolina Wood Duck, *Aix sponsa* (Morse and Wight, 1969), and Goldeneye, *Bucephala clangula* (Andersson and Eriksson, 1982), and in a variety of other species reviewed by Yom-Tov (1980) and Payne (1977). It is more difficult to detect than interspecific nest parasitism such as by cuckoos and others. In general, intraspecific egg dumping is disadvantageous to the host female who owns the nest, who may respond by deserting it if the dumping is excessive, or by reducing the number of eggs she lays herself (Andersson, 1983). None of these species appears to be able to discriminate against the dumped eggs in the way that major female ostriches clearly can. But as with ostriches, conditions favouring the evolution of egg dumping include a shortage of nest sites, a high rate of nest destruction and a large clutch size.

In an increasing number of species, it is being shown that birds other than the main resident pair are providing help at the nest. Examples were scarce a generation ago, but are now abundant (see, for example, Brown, 1978; Emlen, 1984; Fry, 1977; Grimes, 1976; Rowley, 1976; Woolfenden, 1976). In a few cases, it has been shown that two or more females are contributing eggs to the same nest and cooperating in rearing them.

Cooperative breeding of this latter type has been particularly well demonstrated in Pukekos, *Porphyrio p. melanotus* (Craig, 1980), and in Groove-billed Anis, *Crotophaga sulcirostris* (Vehrencamp, 1978). Communally laying females of the latter species have been shown to evict selectively the eggs of their companions, but they do so not by discriminating among eggs but by evicting only eggs laid before they themselves started laying. There are a number of different evolutionary routes by which cooperative breeding has been arrived at in different species, and these have been reviewed by Brown (1987) and Emlen (1984); kinship among helpers often seems to be important, as does better feeding of the young, and group defence of a territory, none of which is significant in the ostrich system.

The merging of broods of young occurs in a small number of precocial species, including Shelduck, *Tadorna tadorna* (Williams, 1974), and Common Eider, *Somateria mollissima* (Munro and Bedard, 1977); it was shown that the merging of broods into a crèche sometimes reduces the amount of predation on the ducklings, as was postulated to be the case with ostriches. One suspects that with all these other species in which crèches occur, and with ostriches, there may be much more going on than we currently understand. For example, who can recognize whom in a crèche, and how? Do the parents benefit by being the individuals caring for the crèche, or they are lumbered with it, perhaps because they have the greatest genetic stake in it? Does the protection from predation come only from the dilution effect, or are some offspring obliged to occupy more vulnerable 'buffer' locations at the edge of the crèche? If so, the situation would be analogous to the behaviour of female ostriches in expelling eggs other than their own.

9.9   Conclusions

In summary, the distinctiveness among bird species of the ostrich communal nesting system lies in the ready acceptance of responsibility by one female for the eggs and young of other females. Kin selection plays no part in this apparent altruism. Rather, there is little altruism involved, because there is no cost to the major female because of her ability to discriminate among eggs in favour of her own. The underlying factors determining the evolution and maintenance of the ostrich system appear to be the birds' very large size and their avoidance of predation. They drive the system and fix the main directions in which natural selection operates within the communal nesting system.

The system clearly results from different individual birds adopting different strategies which are in their own individual interests. It has been possible to a fair extent so far to quantify the relative reproductive advantages of different strategies. To some degree, the pay-off from a particular strategy depends on the strategies that others are adopting. And the behaviour of one bird can affect the behaviour of another, in ways that may well not be to its advantage – such as the conflict of interest between a male and female over when to start incubation. Cooperation, competition and manipulation all take place.

Some individuals are more successful than others, as a result mainly of adopting different strategies. There are many inequalities among individuals, and these may make it more or less advantageous to adopt a particular strategy at a given time; such inequalities include not only differences in size, strength or age, but also differences due to past actions, such as prior investment in a nest or number of eggs left to lay. And unpredictable chance effects, such as whether or when jackals may discover a nest, play a large part.

Much more remains to be discovered about the system. Much of it will not be at all easy or practicable to discover in the wild. But we need to determine more fully for any area which females lay how many eggs, in which nests, when, by whom they were fathered, and what became of them and why. We need more information on the factors influencing mate choice and the choice of becoming a major hen. We have to recognize that an individual bird may be breeding for as long as a research worker's career, and that it is its whole lifetime's reproductive success that will be weighted up by natural selection. There is ample scope for many fascinating detailed investigations into such topics as incubation conditions, mechanisms of synchrony, display functions, sperm competition, causes and effects of different egg sizes, and so on, almost indefinitely. And at the same time, one must not forget to marvel at what a remarkable organism the ostrich is, in many aspects of its life, quite apart from its communal nesting system.

Left footprint

References

Adamson, G.A.G. (1964). Observations on the ostrich (*Struthio camelus massaicus* Neumann). *East African Wildlife Journal* **2**: 164.

Andersson, C.J. (1856). *Lake Ngami or Explorations and Discovery During Four Years Wanderings in the Wilds of South Western Africa*. Hurst and Beckett: London.

Andersson, M. (1983). Brood parasitism within species. In: Barnard, C. (ed), *Producers and Scroungers: Strategies of Exploitation and Parasitism*. Croom Helm: London.

Andersson, M. and Eriksson, M.O.G. (1982). Nest parasitism in Goldeneyes *Bucephala clangula*: Some evolutionary aspects. *American Naturalist* **120**: 1–16.

Bannerman, D.A. (1930). *The Birds of Tropical West Africa* Vol. 1. Crown Agents: London.

Bertram, B.C.R. (1978). Living in groups: predators and prey. In: Krebs, J.R. and Davies, N.B. (eds), *Behavioural Ecology: An Evolutionary Approach*, pp. 64–96. Blackwell: Oxford.

Bertram, B.C.R. (1979). Ostriches recognise their own eggs and discard others. *Nature* **279**: 233–234.

Bertram, B.C.R. (1980). Vigilance and group size in ostriches. *Animal Behaviour* **28**: 278–286.

Bertram, B.C.R. and Burger, A.E. (1981a). Aspects of incubation in ostriches. *Ostrich* **52**: 36–43.

Bertram, B.C.R. and Burger, A.E. (1981b). Are Ostrich *Struthio camelus* eggs the wrong colour? *Ibis* **123**: 207–210.

Bleek, W.H.I. and Lloyd, L.C. (1911). *Specimens of Bushman Folklore*. George Allen: London.

Boswall, J. (1977a). Tool-using by birds and related behaviour. *Avicultural Magazine* **83**: 88–97, 146–159, 220–228.

Boswall, J. (1977b). Notes on tool-using by Egyptian vultures. *Bulletin of the British Ornithologists' Club* **97**: 77–78.

Brodkorb, P. (1963). Catalogue of fossil birds. Part 1. *Bulletin of the Florida State Museum, Biological Sciences* **7**(4): 179–293.

Brooke, R.K. (1979). Tool-using by the Egyptian vulture to the detriment of the Ostrich. *Ostrich* **50**: 119–120.

Brown, J.L. (1978). Avian communal breeding systems. *Annual Review of Ecology and Systematics* **9**: 123–155.

Brown, J.L. (1987). *Helping and Communal Breeding in Birds*. Monographs in Behaviour and Ecology. Princeton University Press: Princeton, N.J.

Brown, L.H., Urban, E.K. and Newman, K. (1982). *The Birds of Africa*, Vol. I. Academic Press: London.

Bruning, D.F. (1973). The Greater Rhea chick and egg delivery route. *Natural History* **82**: 68–75.

Bruning, D.F. (1974). Social structure and reproductive behavior in the Greater Rhea. *The Living Bird* **13**: 251–294.

Bruning, D.F. (1991). Ratites. In: Brooke, M. and Birkhead, T. (eds), *The Cambridge Encyclopedia of Ornithology*, pp. 85–88. Cambridge University Press: Cambridge.

Cloudsley-Thompson, J.L. and Mohamed, E.R.M. (1967). Water economy of the ostrich. *Nature* **216**: 1040.

Clutton-Brock, T.H. and Harvey, P.H. (1978). Mammals, resources and reproductive strategies. *Nature* **273**: 191–195.

Cobb, S.M. (1976). The distribution and abundance of the large mammal community of Tsavo National Park, Kenya. D.Phil. thesis, Oxford University.

Cracraft, J. (1974). Phylogeny and evolution of the ratite birds. *Ibis* **116**: 494–521.

Cracraft, J. (1983). Species concepts and speciation analysis. In: Johnston, R.F. (ed.), *Current Ornithology*, pp. 159–187. Plenum Press: New York.

Craig, J.L. (1980). Pair and group breeding behaviour of a communal gallinule, the Pukeko, *Porphyrio p. melanotus. Animal Behaviour* **28**: 593–603.

Cramp, S. and Simmons, K.E.L. (1980). *The Birds of the Western Palaearctic.* Oxford University Press: Oxford.

Crawford, E.C. and Schmidt-Nielsen, K. (1967). Temperature regulation and evaporative cooling in the ostrich. *American Journal of Physiology* **212**: 347–353.

Crome, F.H.J. (1976). Some observations on the biology of the Cassowary in northern Queensland. *Emu* **76**: 8–14.

Crook, J.H. (1965). The adaptive significance of avian social organisation. *Symposium of the Zoological Society of London* **14**: 181–218.

Davies, S.J.J.F. (1963). Emus. *Australian Natural History* **14**: 225–229.

Davies, S.J.J.F. (1976). The natural history of the emu in comparison with that of the other ratites. *Proceedings of the 16th International Ornithological Congress*, pp. 109–120. Australian Academy of Science: Canberra.

Dawkins, R. (1980). Good strategy or evolutionarily stable strategy? In: Barlow, G.W. and Silverberg, J. (eds) *Sociobiology: Beyond Nature/Nurture*, pp. 331–367. Westview Press: Boulder, Colorado.

De Mosenthal, J. and Harting, J.E. (1879). *Ostriches and Ostrich Farming.* J.C. Juta: Cape Town.

Drent, R.H. (1975). Incubation. In: Farner, D.S. and King, J.R. (eds), *Avian Biology*, Vol. 5, pp. 333–420. Academic Press: New York.

Emlen, S.T. (1984). Co-operative breeding in birds and mammals. In: Krebs, J.R. and Davies, N.B. (eds), *Behavioural Ecology: An Evolutionary Approach*, pp. 305–335. Blackwell: Oxford.

Emlen, S.T. and Oring, L.W. (1977). Ecology, sexual selection, and the evolution of mating systems. *Science* **197**: 215–223.

Fedak, M.A. and Seeherman, H.J. (1979). Reappraisal of energetics of locomotion shows identical cost in bipeds and quadrupeds including ostrich and horse. *Nature* **282**: 713–716.

Foster, J.B. and Coe, M.J. (1968). The biomass of game animals in Nairobi National Park, 1960–66. *Journal of Zoology* **155**: 413–425.

Fry, C.H. (1977). The evolutionary significance of co-operative breeding in birds. In: Stonehouse, B. and Perrins, C.M. (eds), *Evolutionary Ecology*, pp. 127–136. Macmillan: London.

Gaukrodger, D.W. (1925). The emu at home. *Emu* **25**: 53–57.

Geist, V. (1977). A comparison of social adaptations in relation to ecology in gallinaceous birds and ungulate societies. *Annual Review of Ecology and Systematics* **8**: 193–207.

Grimes, L.G. (1976). The occurrence of co-operative breeding behaviour in African birds. *Ostrich* **47**: 1–15.

Hamilton, W.D. (1964). The genetical evolution of social behaviour, I and II. *Journal of Theoretical Biology* **7**: 1–16, 17–52.

Handford, P. and Mares, M.A. (1985). The mating systems of ratites and tinamous: An evolutionary perspective. *Biological Journal of the Linnean Society* **25**: 77–104,

Hoyt, D.F., Vleck, D. and Vleck, C.M. (1978). Metabolism of avian embryos: Ontogeny and temperature effects in the ostrich. *Condor* **80**: 265–271.

Hurxthal, L.M. (1979). Breeding behaviour of the Ostrich *Struthio camelus massaicus* Neumann in Nairobi Park. Ph.D. thesis, Nairobi University.

Jarman, P.J. (1974). The social organisation of antelope in relation to their ecology. *Behaviour* **48**: 215–267.

Jarvis, M.J.F., Jarvis, C. and Keffen, R.H. (1985). Breeding seasons and laying patterns of the southern African Ostrich *Struthio camelus*. *Ibis* **127**: 442–449.

Jenni, D.A. (1974). Evolution of polyandry in birds. *American Zoologist* **14**: 129–144.

Kruuk, H. (1972). *The Spotted Hyena: A Study of Predation and Social Behavior*. Chicago University Press: Chicago, Ill.

Lack, D. (1968). *Ecological Adaptations for Breeding in Birds*. Methuen: London.

Lancaster, D.A. (1964). Life history of the Boucard Tinamou in British Honduras. Part 1: Distribution and general behavior; Part 2: Breeding biology. *Condor* **66**: 165–181, 253–276.

Laufer, B. (1926). *Ostrich Egg-shell Cups of Mesopotamia and the Ostrich in Ancient and Modern Times*. Leaflet No. 23. Field Museum of Natural History: Chicago, Ill.

Leuthold, W. (1970). Some breeding data on Somali Ostrich. *East African Wildlife Journal* **8**: 206.

Leuthold, W. (1977). Notes on the breeding biology of the ostrich *Struthio camelus* in Tsavo East National Park, Kenya. *Ibis* **119**: 541–544.

Louw, G.N. (1972). The role of advective fog in the water economy of certain Namib Desert animals. In: Maloiy, G.M.O. (ed.), *Comparative Physiology of Desert Animals*. pp. 297–314. Academic Press: London.

Louw, G.N., Belonje, P.C. and Coetzee, H.J. (1969). Renal function, respiration, heart rate and thermoregulation in the Ostrich (*Struthio camelus*). *Scientific Papers of the Namib Desert Research Station* **42**: 43–54.

Lundy, H. (1969). A review of the effects of temperature, humidity, turning and gaseous environment on the hatchability of the hen's egg. In: Carter, T.C. and Freeman, B.M. (eds), *The Fertility and Hatchability of the Hen's Egg*, pp. 143–176. Oliver and Boyd: Edinburgh.

Mackworth-Praed, C.W. and Grant, C.H.B. (1952). *Birds of Eastern and North Eastern Africa*. Longmans, Green and Co.: London.

Mackworth-Praed, C.W. and Grant, C.H.B. (1962). *Birds of the Southern Third of Africa*. Longmans: London.

Maynard-Smith, J. (1976). Evolution and the theory of games. *American Scientist* **64**: 41–45.

Mitchell, E.K. (1960). *The Ostrich and Ostrich Farming*. Bibliographical Series of the University of Cape Town School of Librarianship: Cape Town.

Morse, T.E. and Wight, H.M. (1969). Dump nesting and its effect on production in wood ducks. *Journal of Wildlife Management* **33**: 284–293.

Munro, J. and Bedard, J. (1977). Gull predation and creching behaviour in the Common Eider. *Journal of Animal Ecology* **46**: 799–810.

Orians, G.H. (1969). On the evolution of mating systems in birds and mammals. *American Naturalist* **103**: 589–603.

Payne, R.B. (1977). The ecology of brood parasitism in birds. *Annual Review of Ecology and Systematics* **8**: 1–28.

Pearson, A.K. and Pearson, O.P. (1955). Natural history and breeding behavior of the tinamou, *Nothoprocta ornata*. *Auk* **72**: 113–127.

Pienaar, U. de V. (1969). Predator–prey relations amongst the larger mammals of the Kruger National Park. *Koedoe* **9**: 40–107.

Rahn, H. and Ar, A. (1974). The avian egg: Incubation time and water loss. *Condor* **76**: 147–152.

Rahn, H., Paganelli, C.V. and Ar, A. (1975). Relation of avian egg weight to body weight. *Auk* **92**: 750–765.

Reid, B. and Williams, G.R. (1975). The Kiwi. In: Kuschel, G. (ed.), *Biogeography and Ecology in New Zealand*, pp. 301–330. W. Junk: The Hague.

Rowley, I. (1976). Co-operative breeding in Australian birds. *Proceedings of the 16th International Ornithological Congress*, pp. 657–666. Australian Academy of Science: Canberra.

Rudnai, J. (1974). The pattern of lion predation in Nairobi Park. *East African Wildlife Journal* **12**: 213–225.

Sauer, E.G.F. (1970). Interspecific behaviour of the South African Ostrich. *Ostrich* **8**: 91–103 (suppl.).

Sauer, E.G.F. (1971). Zur Biologie der wilden Strausse Sudwestafrikas. *Zeitschrift des Kolner Zoo* **14**: 43–64.

Sauer, E.G.F. (1982). Aberrant sexual behavior in the South African Ostrich. *Auk* **89**: 717–737.

Sauer, E.G.F. and Sauer, E.M. (1966a). The behavior and ecology of the South African Ostrich. *The Living Bird* **5**: 45–75.

Sauer, E.G.F. and Sauer, E.M. (1966b). Social behaviour of the South African Ostrich, *Struthio camelus australis*. *Ostrich* **6**: 183–191 (suppl.).

Sauer, E.G.F. and Sauer, E.M. (1967). Yawning and other maintenance activities in the South African Ostrich. *Auk* **84**: 571–587.

Schafer, E. (1954). Zur Biologie des Steisshuhnes *Nothocercus bonpartei*. *Journal of Ornithology* **95**: 219–232.

Schaller, G.B. (1972). *The Serengeti Lion*. Chicago University Press: Chicago, Ill.

Sibley, C.G. and Ahlquist, J.E. (1981). The phylogeny and relationships of the ratite birds as indicated by DNA–DNA hybridisation. *Evolution Today: Proceedings of the 2nd International Congress of Systematics and Evolutionary Biology*, pp. 301–335 The Hunt Institute for Botanical Documentation: Pittsburgh, USA.

Siegfried, W.R. and Frost, P.G.H. (1974). Egg temperature and incubation behaviour of the ostrich. *Madoqua* **8**: 63–66.

Smit, D.J.v.Z. (1963) *Ostrich Farming in the Little Karoo*. Department of Agricultural Technical Services Bulletin No. 358. Government Printer: Pretoria.

Swart, D. (1978). Opbergingsperiode van volstruiseiers. *Elsenburg Joernaal* **2**(2): 19–20.

Taylor, C.R., Schmidt-Nielsen, K. and Raab, J.L. (1970). Scaling energetic cost of running to body weight of mammals. *American Journal of Physiology* **219**: 1104–1107.

Thouless, C.R., Fanshawe, J.H. and Bertram, B.C.R. (1989). Egyptian vultures *Neophron percnopterus* and Ostrich *Struthio camelus* eggs: The origins of stone-throwing behaviour. *Ibis* **131**: 9–15.

Van Lawick-Goodall, J. (1968). Tool-using bird: The Egyptian vulture. *National Geographic* **133**: 630–641.

Van Rhijn, J. (1984). Phylogenetical constraints in the evolution of parental care strategies in birds. *Netherlands Journal of Zoology* **34**: 103–122.

Vehrencamp, S.L. (1978). The adaptive significance of communal nesting in Groove-billed Anis (*Crotophaga sulcirostris*). *Behavioral Ecology and Sociobiology* **4**: 1–33.

Vince, M.A. (1969). Embryonic communication, respiration and the synchronisation of hatching. In: Hinde, R.A. (ed.), *Bird Vocalisations*, pp. 233–260. Cambridge University Press: Cambridge.

Western, D. (1973). The structure, dynamics, and changes of the Amboseli ecosystem. Ph.D. thesis, University of Nairobi.

Williams, G.C. (1966). *Adaptation and Natural Selection*. Princeton University Press: Princeton, N.J.

Williams, M.J. (1974). Creching behaviour of the Shelduck *Tadorna tadorna* L. *Ornis Scandinavica* **5**: 131–143.

Woolfenden, G.E. (1976). Co-operative breeding in American birds. *Proceedings of the 16th International Ornithological Congress*, pp. 674–684. Australian Academy of Science: Canberra.

Yom-Tov, Y. (1980). Intraspecific nest parasitism in birds. *Biological Reviews* **55**: 93–108.

Subject Index

activity patterns: before nesting, 49–50; categories recorded, 22–24; of major and minor females, 134–135; of males around nesting, 140–141
age structure, 33–34
altruism, 17, 186
anatomy, 1–2, 12
Apteryx, see Kiwi
attendance, *see* nest attendance

behaviour patterns, 14–16
Black-backed jackal, *see* jackal
body size, 1; and egg size, 175–177; as factor in evolution of communal nesting system, 173–177; in relation to major or minor strategies, 135
booming, *see* singing
brain weight: and body size, 177
breakage of eggs: extent of, 64–66; in surviving nests, 107–108

calcium: requirement for egg shells, 120–121
Canis mesomelas, see jackal
Cassowary, 3, 181–185
Casuarius, see Cassowary
chicks: definition of, 33; observation of, 68–69; survival of, 89–90; *see also* crèching
choice of partner, 160–162
classification, 2–3, 12
climate: effect on nesting season, 95–96; as factor in evolution of communal nesting system, 179–180; in Tsavo West National Park, 20–21; *see also* temperature of environment
colour of eggs: effect on predation risk, 87–88; effect on temperature, 88–89
communal breeding: in other species, 16–17, 185–186

conspicuousness: of eggs, 87–89, 175; of females, 172; of groups, 37; of males, 36–38, 46, 172–173; of nest after fire, 144
copulation, *see* mating
courtship, *see* kantling
crèching: description of, 68–69; in other species, 186
Crocuta crocuta, see hyaenas

deception: of males by females, 162
detection distances, 37–39
detection of birds, 22, 37
dilution: of predation on nests, 107–109, 120; of predator attack on birds, 90, 93, 95, 120, 186
displays: distraction display, 15–16, 76; soliciting, 15, 23, 26, 57, 60; threat, 15; wing flagging, 15, 45; *see also* kantling
distribution, 42–48; *see also* territory; ranges
domestication, *see* farming
dominance: between major and minor females, 136, 182
Dromaius novaehollandiae, see Emu
dummy nests, 29, 80–83

egg discrimination: benefits of, 118–119; cues used in, 111–118, 165; by major female, 110–120; and optimum time to start incubation, 169; *see also* outer ring
egg dumping, *see* nest parasitism
egg shape, 115, 117
egg shell: use by humans, 8–9, 11; thickness of, 175
egg size, 2; changes with laying order, 132; consistency within individuals in, 113–116; evolutionary effects of, 175–177; of major versus minor females, 132; measurement of, 30; in relation to number laid, 130–131; in relation to shell